Series Editor

BEYOND THE COMMON CORE

A HANDBOOK FOR

Mathematics

in a PLC at Work™

GRADES 6-8

Solution Tree | Press

a division of

Solution Tree

A Joint Publication With

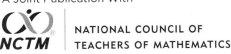
NATIONAL COUNCIL OF TEACHERS OF MATHEMATICS

Jessica Kanold-McIntyre
Matthew R. Larson
Diane J. Briars

555 North Morton Street
Bloomington, IN 47404
800.733.6786 (toll free) / 812.336.7700
FAX: 812.336.7790

email: info@solution-tree.com
solution-tree.com

Visit **go.solution-tree.com/mathematicsatwork** to download the reproducibles in this book.

Printed in the United States of America

19 18 17 16 15   1 2 3 4 5

Library of Congress Cataloging-in-Publication Data

Kanold-McIntyre, Jessica, author.
   Beyond the common core : a handbook for mathematics in a PLC at work. Grades 6-8 / Jessica Kanold-McIntyre, Matthew R. Larson, Diane J. Briars.
      pages cm. -- (Beyond the common core)
   Includes bibliographical references and index.
   ISBN 978-1-936763-48-1 (perfect bound) 1. Mathematics--Study and teaching (Middle school)--United States.
2. Professional learning communities. I. Larson, Matthew R., author. II. Briars, Diane J., 1951- author. III. Title.
IV. Title: Handbook for mathematics in a PLC at work. Grades 6-8.
   QA135.6.K195 2015
   510.71'273--dc23
                                        2014041635

Solution Tree
Jeffrey C. Jones, CEO
Edmund M. Ackerman, President

Solution Tree Press
President: Douglas M. Rife
Associate Acquisitions Editor: Kari Gillesse
Editorial Director: Lesley Bolton
Managing Production Editor: Caroline Weiss
Senior Production Editor: Suzanne Kraszewski
Copy Editor: Sarah Payne-Mills
Proofreader: Elisabeth Abrams
Text and Cover Designer: Laura Kagemann
Compositor: Rachel Smith

Acknowledgments

I would like to thank Solution Tree for the great opportunity to see a dream become a reality. Thank you to the vision of the writing team of Tim, Diane, and Matt: I have learned so much from the three of you. Most importantly, I would like to acknowledge the amazing teachers and leaders at Aptakisic-Tripp CCSD 102 for their ability to embrace the challenge of implementing the Common Core State Standards in Mathematics while also keeping student learning as the primary focus. Special thanks to Amy Leibach, David Lee, Kelly Morales, Adam Rosenak, Jacqui Giuliano, and Nicole Schneider—you are my heroes! Personally, I would like to thank my dad, Tim Kanold, for pushing my own belief in myself beyond what I could imagine. Finally, thank you to my husband, Tim, for his constant support and patience through this writing process and my own professional journey.

— Jessica Kanold-McIntyre

Once again, I find myself thanking Tim Kanold for his vision, positive perseverance, and sincere commitment to improving mathematics education in the United States. Your leadership, mentorship, and friendship are deeply appreciated. A sincere thank you to the leaders and teachers of mathematics in the Lincoln Public Schools, particularly Delise Andrews, Susie Katt, Julie Kreizel, and Jerel Welker, who over the last two decades have provided critical feedback on the recommendations in this handbook and challenged me as we collaborated to improve student learning. Finally, I owe a special thanks to Tammy, my wife, who always supports and encourages my professional endeavors, often at her own personal sacrifice.

—Matthew R. Larson

First, thanks to Tim Kanold for his vision about the critical role of PLCs in systemic improvement of mathematics teaching and learning for all teachers and students and for tools that effectively support making this vision a reality. I deeply value our friendship and collaboration and have learned much from you over the years. Second, to Jessica and Matt for their insights about teaching and learning in the middle grades that are reflected in this book. And, finally, to Jim, my husband, for his loving support and intellectual challenges that enhance my work every day.

—Diane J. Briars

First and foremost, I extend my thanks to Jessica, Matt, and Diane for understanding the joy, the pain, and the hard work of the writing journey, and for giving freely of their talent to so many others. You are gifted colleagues. My thanks to our reviewers, who have dedicated their lives to the work and effort described within the pages of this handbook.

Thanks, too, to Jeff Jones, Douglas Rife, Lesley Bolton, Suzanne Kraszewski, and Sarah Payne-Mills from Solution Tree for their belief in and dedication to our vision and work in mathematics education. And to Rick and Becky DuFour for their wisdom, insight, advice, and deep understanding of what it takes to become an authentic professional learning community.

Sincere thanks also to our colleagues from the National Council of Teachers of Mathematics (NCTM) and the Educational Materials Committee for their support of this series and their leadership in mathematics education.

Finally, thanks to my wife, Susan, my loving critic, who understands how to formatively guide me through a handbook series project as bold as this.

—Timothy D. Kanold

Solution Tree Press would like to thank the following reviewers.

Ellen Barger
Assistant Superintendent
Santa Barbara County Education Office
Santa Barbara, California

Laura Godfrey
Middle School Math Specialist
Madison Metropolitan School District
Madison, Wisconsin

Beth Kobett
Assistant Professor of Education
Stevenson University
Owings Mills, Maryland

Suzi Mast
Director of K–12 Mathematics
Arizona Department of Education
Phoenix, Arizona

Kelly Morales
Math Teacher
Aptakisic Junior High School
Buffalo Grove, Illinois

Kit Norris
Senior Associate, Common Core PLC Learning Group
Boston, Massachusetts

Sue Pippen
Pippen Consulting
Past President, Illinois Council of Teachers of Mathematics
Plainfield, Illinois

Sarah Schuhl
PLC Associate and Mathematics Educational Consultant
Portland, Oregon

Pamela Richards
Teacher on Special Assignment, Secondary Mathematics & Science
Visalia Unified School District
Visalia, California

Table of Contents

Visit **go.solution-tree.com/mathematicsatwork** to download the reproducibles in this book.

About the Series Editor

Timothy D. Kanold, PhD, is an award-winning educator, author, and consultant. He is former director of mathematics and science and served as superintendent of Adlai E. Stevenson High School District 125, a model professional learning community district in Lincolnshire, Illinois. He serves as an adjunct faculty member for the graduate school at Loyola University Chicago.

Dr. Kanold is committed to a vision for Mathematics at Work™, a process of learning and working together that builds knowledge sharing, equity, and excellence for all students, faculty, and school administrators. He conducts highly motivational professional development leadership seminars worldwide with a focus on turning school vision into realized action that creates increased learning opportunities for students through the effective delivery of professional learning communities for faculty and administrators.

He is a past president of the National Council of Supervisors of Mathematics and coauthor of several best-selling mathematics textbooks. He has served on writing commissions for the National Council of Teachers of Mathematics and the National Council of Supervisors of Mathematics. He has authored numerous articles and chapters on school mathematics, leadership, and professional development for education publications.

In 2010, Dr. Kanold received the prestigious international Damen Award for outstanding contributions to the leadership field of education from Loyola University Chicago. He also received the Outstanding Administrator Award from the Illinois State Board of Education in 1994 and the Presidential Award for Excellence in Mathematics and Science Teaching in 1986.

Dr. Kanold earned a bachelor's degree in education and a master's degree in applied mathematics from Illinois State University. He completed a master's in educational administration at the University of Illinois and received a doctorate in educational leadership and counseling psychology from Loyola University Chicago.

To learn more about Dr. Kanold's work, visit his blog *Turning Vision Into Action* at http://tkanold.blogspot.com, or follow @tkanold on Twitter.

To book Dr. Kanold for professional development, contact pd@solution-tree.com.

About the Authors

Jessica Kanold-McIntyre is principal of Aptakisic Junior High School in Buffalo Grove, Illinois. She oversees her teachers in the implementation of initiatives such as the Common Core; Next Generation Science Standards; and the Social Studies College, Career, and Civic Life framework, while also supporting a 1:1 iPad environment for students. She focuses teacher instruction through the professional learning community (PLC) process, creating learning opportunities around formative assessment practices and student engagement.

Jessica is also the Aptakisic-Tripp Community Consolidated School District 102 mathematics leader. As principal, Jessica has developed and implemented a districtwide process for the Common Core State Standards and has helped to create and implement curriculum guides for K–8 mathematics, algebra 1, and algebra 2. She previously served as assistant principal at Aptakisic, where she led and supported special education, response to intervention (RTI), and English learner staff through the PLC process.

Jessica's teaching experience includes serving as a mathematics teacher for traditional and honors students. She was the pilot teacher for Promethean Interactive whiteboard technology in her district, helping to develop a 21st century classroom. As an educator and leader, Jessica is committed to providing students with cutting-edge 21st century experiences that engage and challenge them.

She earned a bachelor's degree in elementary education from Wheaton College and a master's degree in educational administration from Northern Illinois University.

Matthew R. Larson, PhD, an award-winning educator and author, is the K–12 mathematics curriculum specialist for Lincoln Public Schools in Nebraska. He served on the National Council of Teachers of Mathematics board of directors, is currently president elect of NCTM, and has served as a member of the Institute of Education Sciences What Works Clearinghouse panel on improving algebra skills. Dr. Larson has taught mathematics at the elementary through college levels and has held an honorary appointment as a visiting associate professor of mathematics education at Teachers College, Columbia University.

He is coauthor of several mathematics textbooks, professional books, and articles in mathematics education. A frequent keynote speaker at national meetings, Dr. Larson's humorous presentations are well-known for their application of research findings to practice.

Dr. Larson earned a bachelor's degree and doctorate from the University of Nebraska–Lincoln.

Diane J. Briars, PhD, a mathematics education consultant, is a president of the National Council of Teachers of Mathematics. Dr. Briars was mathematics director of Pittsburgh Public Schools for twenty years. Under her leadership, Pittsburgh schools made significant progress in increasing student achievement through standards-based curricula, instruction, and assessment. She is past president of the National Council of Supervisors of Mathematics and senior developer and research associate for the Algebra Intensification Project. Dr. Briars began her career as a secondary mathematics teacher.

Dr. Briars has been a member of many committees, including the National Commission on Mathematics and Science Teaching for the 21st Century. She has served in leadership roles for various national organizations, including the National Council of Teachers of Mathematics, the College Board, and the National Science Foundation.

She earned a doctorate in mathematics education and a master's and bachelor's in mathematics from Northwestern University.

To book Jessica Kanold-McIntyre, Dr. Larson, Dr. Briars, or Dr. Kanold for professional development, contact pd@solution-tree.com.

Introduction

You have high impact on the front lines as you snag children in the river of life.

—Tracy Kidder

Your work as a middle school mathematics teacher is one of the most important, and at the same time, one of the most difficult jobs to do well in education. Since the release of our 2012 Solution Tree Press series *Common Core Mathematics in a PLC at Work*™, our authors, reviewers, school leaders, and consultants from the Mathematics at Work™ team have had the opportunity to work with thousands of grade 6–8 teachers and teacher teams from across the United States who are just like you: educators trying to urgently and consistently seek deeper and more meaningful solutions to a sustained effort for meeting the challenge of improved student learning in mathematics. From California to Virginia, Utah to Florida, Oregon to New York, Wisconsin to Texas, and beyond, we have discovered a thirst for implementation of K–12 mathematics programs that will sustain student success over time. Your focus on middle school mathematics is one of the most significant components of the K–12 effort toward improved student learning.

Certainly, the Common Core State Standards (CCSS) have served as a catalyst for much of the national focus and conversation about improving student learning. However, your essential work as a middle school mathematics teacher and as part of a collaborative team in your local school and district takes you well *beyond* your state's standards—whatever they may be. As the authors of the National Council of Teachers of Mathematics (NCTM, 2014) publication *Principles to Actions: Ensuring Mathematical Success for All* argue, standards in and of themselves do not describe the essential conditions necessary to ensure mathematics learning for all students. You, as the classroom teacher, are the most important ingredient to student success.

This middle school mathematics teaching and assessing handbook is designed to take you *beyond the product* of standards themselves by providing you and your collaborative team with the guidance, support, and *process* tools necessary to achieve mathematics program and department greatness within the context of higher levels of demonstrated student learning and performance.

Whether you are from a state that is participating in one of the CCSS assessment consortia or from a state that uses a unique mathematics assessment designed only for your state, it is our hope that this handbook provides a continual process that allows you to move toward a local program of great mathematics teaching and learning for you and your students.

Your daily work begins by understanding there are thousands of instructional and assessment decisions you and your teacher team (those adults closest to the action) will make every day and in every unit. Do those decisions make a significant difference in terms of increased levels of student achievement? Your role as a middle school teacher is to make sure they do.

The Grain Size of Change Is the Teacher Team

We believe the best strategy to achieve the expectations of CCSS, state, or local standards for mathematics is to create schools and districts that operate as professional learning communities (PLCs), and, more specifically, within a PLC at Work™ culture as outlined by Richard DuFour, Rebecca DuFour, Robert Eaker, and Tom Many (2010). We believe that the PLC process supports a grain size of change that is just right—not too small (the individual teacher) and not too big (the district office)—for impacting deep change. The adult knowledge capacity development and growth necessary to deliver on the promise of standards that expect student demonstrations of understanding reside in the engine that drives the PLC school culture: you and your teacher team.

There is a never-ending aspect to your professional journey and the high-leverage teacher and teacher team actions that measure your impact on student learning. This idea is at the very heart of your work. As John Hattie (2012) states in *Visible Learning for Teachers: Maximizing Impact on Learning*:

> My role as a teacher is to evaluate the effect I have on my students. It is to "know thy impact," it is to understand this impact, and it is to act on this knowing and understanding. This requires that teachers gather defensible and dependable evidence from many sources, and hold collaborative discussions with colleagues and students about this evidence, thus making the effect of their teaching visible to themselves and to others. (p. 19)

Knowing Your Vision for Mathematics Instruction and Assessment

Quick—you have thirty seconds: turn to a colleague and declare your vision for mathematics instruction and assessment in your middle school mathematics department for your school. What exactly will you say? More importantly, on a scale of 1 (low) to 6 (high), what would be the degree of coherence between your and your colleagues' visions for instruction and assessment?

We have asked these vision questions to more than ten thousand mathematics teachers across the United States since 2011, and the answers have been consistent: wide variance on mathematics instruction and assessment coherence from teacher to teacher (low scores of 1, 2, or 3 mostly) and general agreement that the idea of some type of a formative assessment process is supposed to be in a vision for mathematics instruction and assessment.

A favorite team exercise we use to capture the vision for instruction and assessment is to ask a team of three to five teachers to draw a circle in the middle of a sheet of poster paper. We ask each team member to write a list (outside of the circle) of three or four vital adult behaviors that reflect his or her vision for instruction and assessment. After brainstorming, the team will have twelve to fifteen vital teacher behaviors.

We then ask the team to prepare its vision for mathematics instruction and assessment inside the circle. The vision must represent the vital behaviors each team member has listed in eighteen words or less. We indicate, too, that the vision should describe a "compelling picture of the school's future that produces energy, passion, and action in yourself and others" (Kanold, 2011, p. 12).

Team members are allowed to use pictures, phrases, or complete sentences, but all together, the vision cannot be more than eighteen words. In almost every case, in all of our workshops, professional development events, conferences, institutes, and onsite work, we have been asked a simple, yet complex question: "*How?*" How do you begin to make decisions and do your work in ways that will advance your vision for mathematics instruction and assessment in your middle school? How do you honor what is inside your circle? And how do you know that your circle, your defined vision for mathematics instruction and assessment, represents the "right things" to pursue that are worthy of your best energy and effort?

In *Common Core Mathematics in a PLC at Work, Grades 6–8* (Kanold, Briars, Asturias, Foster, & Gale, 2013), we explain how understanding *formative assessment* as a research-affirmed *process* for student and adult learning serves as a catalyst for successful mathematics content implementation. In the *Common Core Mathematics in a PLC at Work* series, we establish the pursuit of assessment as a process of formative feedback and learning for the students and the adults as a highly effective practice to pursue (see chapter 4 in any book of the series).

In this handbook, we provide tools for *how* to achieve that collaborative pursuit: how to engage in ten *high-leverage team actions* (HLTAs) steeped in a commitment to a vision for mathematics instruction and assessment that will result in greater student learning than ever before.

A Cycle for Analysis and Learning: The Instructional Unit

The mathematics unit or chapter of content creates a natural cycle of manageable time for a teacher's and team's work throughout the year. What is a *unit*? For the purposes of your work in this handbook, we define a *unit* as a chunk of mathematics content. It might be a chapter from your textbook or other materials for the course, a part of a chapter or set of materials, or a combination of various short chapters or content materials. A unit generally lasts no less than two to three weeks and no more than four to five weeks.

As DuFour, DuFour, and Eaker (2008), the architects of the PLC at Work process, advise, there are four critical questions every collaborative team in a PLC at Work culture asks and answers on a unit-by-unit basis:

1. What do we want all students to know and be able to do? (The essential learning standards)
2. How will we know if they know it? (The assessment instruments and tasks teams use)
3. How will we respond if they don't know it? (Formative assessment processes for intervention)
4. How will we respond if they do know it? (Formative assessment processes for extension and enrichment)

The unit or chapter of content, then, becomes a natural cycle of time that is not too small (such as one week) and not too big (such as nine weeks) for meaningful analysis, reflection, and action by you and your teacher team throughout the year as you seek to answer the four critical questions of a PLC. A unit should be analyzed based on content standard clusters—that is, three to five essential standards (or sometimes a cluster of standards) for the unit. Thus, a teacher team, an administrative team, or a district office team, does this type of analysis about eight to ten times per year.

This Mathematics at Work handbook consists of three chapters that fit the natural rhythm of your ongoing work as a teacher of mathematics and as part of a teacher team. The chapters bring a focus to ten high-leverage team actions your team takes before, during, and in the immediate aftermath of a unit of instruction as you respond to the four critical questions of a PLC throughout the year, as highlighted in figure I.1 (page 4). Figure I.1 lists the ten high-leverage team actions within their time frame in relation to the unit of instruction (before, during, or after) and then links the actions to the critical questions of a PLC that they address.

High-Leverage Team Actions	1. What do we want all students to know and be able to do?	2. How will we know if they know it?	3. How will we respond if they don't know it?	4. How will we respond if they do know it?
Before-the-Unit Team Actions				
HLTA 1. Making sense of the agreed-on essential learning standards (content and practices) and pacing	▨			
HLTA 2. Identifying higher-level-cognitive-demand mathematical tasks	▨	◧		
HLTA 3. Developing common assessment instruments	◧	▨		
HLTA 4. Developing scoring rubrics and proficiency expectations for the common assessment instruments		▨		
HLTA 5. Planning and using common homework assignments	◧	▨	◧	◧
During-the-Unit Team Actions				
HLTA 6. Using higher-level-cognitive-demand mathematical tasks effectively	◧	▨		
HLTA 7. Using in-class formative assessment processes effectively	◧	◧	▨	▨
HLTA 8. Using a lesson-design process for lesson planning and collective team inquiry	▨	▨	▨	▨
After-the-Unit Team Actions				
HLTA 9. Ensuring evidence-based student goal setting and action for the next unit of study			▨	▨
HLTA 10. Ensuring evidence-based adult goal setting and action for the next unit of study			▨	▨

▨ = Fully addressed with high-leverage team action

◧ = Partially addressed with high-leverage team action

Figure I.1: High-leverage team actions aligned to the four critical questions of a PLC.

Visit **go.solution-tree.com/mathematicsatwork** to download a reproducible version of this figure.

Before the Unit

In chapter 1, we provide insight into the work of your collaborative team *before* the unit begins, along with the tools you will need in this phase. Your collaborative team expectation should be (as best you can) to complete this teaching and assessing work in preparation for the unit.

There are five before-the-unit high-leverage team actions for collaborative team agreement on a unit-by-unit basis.

> HLTA 1. Making sense of the agreed-on essential learning standards (content and practices) and pacing
>
> HLTA 2. Identifying higher-level-cognitive-demand mathematical tasks
>
> HLTA 3. Developing common assessment instruments
>
> HLTA 4. Developing scoring rubrics and proficiency expectations for the common assessment instruments
>
> HLTA 5. Planning and using common homework assignments

Once your team has taken these action steps, the mathematics unit begins.

During the Unit

In chapter 2, we provide the tools for and insight into the formative assessment work of your collaborative team *during* the unit. This chapter teaches deeper understanding of content, discussing the Mathematical Practices and processes and using higher-level-cognitive-demand mathematical tasks effectively. It helps your team with daily lesson design and study ideas as ongoing in-class student assessment becomes part of a teacher-led formative process.

This chapter introduces three during-the-unit high-leverage team actions your team works through on a unit-by-unit basis.

> HLTA 6. Using higher-level-cognitive-demand mathematical tasks effectively
>
> HLTA 7. Using in-class formative assessment processes effectively
>
> HLTA 8. Using a lesson-design process for lesson planning and collective team inquiry

The end of each unit results in some type of student assessment. You pass back the assessments scored and with feedback. Then what? What are students to do? What are you to do?

After the Unit

In chapter 3, we provide tools for and insight into the formative work your collaborative team does *after* the unit is over. After students have taken the common assessment, they are expected to reflect on the results of their work and use the common unit assessment instrument for formative feedback purposes.

In addition, there is another primary formative purpose to using a common end-of-unit assessment, which Hattie (2012) describes in *Visible Learning for Teachers*: "This [teachers collaborating] is not critical reflection, but *critical reflection in light of evidence* about their teaching" (p. 19, emphasis added).

From a practical point of view, an end-of-unit analysis of the common assessment focuses your team's next steps for teaching and assessing for the next unit. Thus, there are two end-of-unit high-leverage team actions your team works through on a unit-by-unit basis.

> HLTA 9.   Ensuring evidence-based student goal setting and action for the next unit of study
>
> HLTA 10. Ensuring evidence-based adult goal setting and action for the next unit of study

In *Principles to Actions: Ensuring Mathematical Success for All*, NCTM (2014) presents a modern-day view of professional development for mathematics teachers. It describes teachers as professionals who continually seek to improve their mathematical knowledge of teaching, knowledge of mathematical pedagogy, and knowledge of students as learners of mathematics through ongoing learning and collaboration with colleagues.

More importantly, however, you and your colleagues can intentionally *act* on that ever-enhancing knowledge base and transfer what you learn into daily classroom practice through the ten high-leverage teacher team actions presented in this handbook. For more information on the connection between these two documents, see appendix E, p. 189.

Although given less attention, the difficult work of collective inquiry and action orientation has a more direct impact on student learning than when you work in isolation (Hattie, 2009). Through your team commitment (the engine that drives the PLC at Work culture and processes of collective inquiry and action research), you will find meaning in the collaborative work with your colleagues.

In *Great by Choice*, Jim Collins (2011) asks, "Do we really believe that our actions count for little, that those who create something great are merely lucky, that our circumstances imprison us?" He then answers, "Our research stands firmly against this view. Greatness is not primarily a matter of circumstance; greatness is first and foremost a matter of conscious choice and discipline" (p. 181). We hope this handbook helps you focus your time, energy, choices, and pursuit of a great teaching journey.

CHAPTER 1

Before the Unit

Teacher: Know thy impact.

—John Hattie

The ultimate outcome of planning before the unit is for you and your team members to gain a clear understanding of the impact of your expectations for student learning and demonstrations of understanding during the unit.

In conjunction with the scope and sequence your district mathematics curriculum provides, your collaborative team prepares a roadmap that describes the knowledge students will know and be able to demonstrate at the conclusion of the unit. To create this roadmap, your collaborative team prepares and organizes your work around five before-the-unit-begins high-leverage team actions.

> HLTA 1. Making sense of the agreed-on essential learning standards (content and practices) and pacing
>
> HLTA 2. Identifying higher-level-cognitive-demand mathematical tasks
>
> HLTA 3. Developing common assessment instruments
>
> HLTA 4. Developing scoring rubrics and proficiency expectations for the common assessment instruments
>
> HLTA 5. Planning and using common homework assignments

These five team pursuits are based on step one of the PLC teaching-assessing-learning cycle (Kanold, Kanold, & Larson, 2012) shown in figure 1.1 (page 8). This cycle drives your pursuit of a meaningful formative assessment and learning process for your team and for your students throughout the unit and the year.

In this chapter, we describe each of the five before-the-unit-begins high-leverage team actions in more detail (the what) along with suggestions for how to achieve these pursuits (the how). Each HLTA section ends with an opportunity for you to evaluate your current reality (your team's progress). The chapter ends with time for reflection and action (setting your Mathematics at Work priorities for team action).

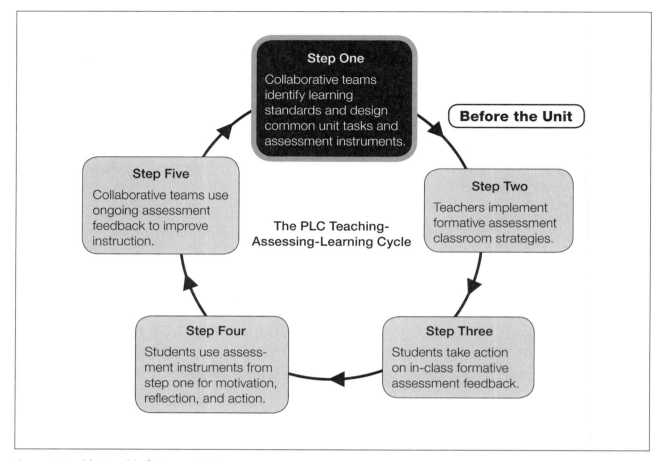

Source: Kanold, Kanold, & Larson, 2012.

Figure 1.1: Step one of the PLC teaching-assessing-learning cycle.

HLTA 1: Making Sense of the Agreed-On Essential Learning Standards (Content and Practices) and Pacing

An excellent mathematics program includes curriculum that develops important mathematics along coherent learning progressions and develops connections among areas of mathematical study and between mathematics and the real world.

—National Council of Teachers of Mathematics

In most middle school mathematics courses, there will be ten to twelve mathematics units (or chapters) during the school year. These units may also consist of several learning modules depending on how your middle school curriculum and courses are structured. An ongoing challenge is for you and your team to determine how to best make sense of and develop understanding for each of the agreed-on essential learning standards within the mathematics unit.

Recall there are four critical questions every collaborative team in a PLC asks and answers on an ongoing unit-by-unit basis.

1. What do we want all students to know and be able to do? (The essential learning standards)
2. How will we know if they know it? (The assessment instruments and tasks teams use)
3. How will we respond if they don't know it? (Formative assessment processes for intervention)
4. How will we respond if they do know it? (Formative assessment processes for extension and enrichment)

High-Leverage Team Action	1. What do we want all students to know and be able to do?	2. How will we know if they know it?	3. How will we respond if they don't know it?	4. How will we respond if they do know it?
Before-the-Unit Action				
HLTA 1. Making sense of the agreed-on essential learning standards (content and practices) and pacing	▢			

▢ = Fully addressed with high-leverage team action

The What

This first high-leverage team action enhances clarity on the first PLC critical question for collaborative team learning: What do we want all students to know and be able to do? In light of the Common Core State Standards for mathematics, the essential learning standards for the unit—the guaranteed and viable mathematics curriculum—include the essential standards students will learn, when they will learn each essential standard (the pacing of the unit), and how they will learn it via process standards such as the Common Core Standards for Mathematical Practice or NCTM's eight teaching practices (2014, p. 10). For example, the Standards for Mathematical Practice "describe varieties of expertise that mathematics educators at all levels should seek to develop in their students" (National Governors Association Center for Best Practices [NGA] & Council of Chief State School Officers [CCSSO], 2010, p. 6). Following are the eight Standards for Mathematical Practice, which we include in full in appendix A (page 177).

1. Make sense of problems and persevere in solving them.
2. Reason abstractly and quantitatively.
3. Construct viable arguments and critique the reasoning of others.
4. Model with mathematics.
5. Use appropriate tools strategically.
6. Attend to precision.
7. Look for and make use of structure.
8. Look for and express regularity in repeated reasoning. (NGA & CCSSO, 2010, pp. 6–8)

While schools and districts use many names for learning standards—*learning goals, learning targets, learning objectives*, and so on—this handbook references the broad mathematical concepts and understandings for the entire unit as *essential learning standards*. The essential learning standards will become the focus for your analysis of student performance during the unit. For more specific lesson-by-lesson daily outcomes, we use *daily learning objectives* or *questions*. We use the terms *learning goals* or *learning targets* to reference the *outcome* for student proficiency in each standard. The daily learning objectives and the tasks and activities representing those objectives help students *understand* the essential learning standards for the unit in order to demonstrate proficiency (outcomes) on those standards. The daily learning objectives articulate for students what they are to learn *that day* and at the same time provide insight for teachers on how to assess students on the essential learning standards at the end of the unit. It is important to keep in mind that your daily learning objectives must maintain the same expectation for developing student understanding and not allow the student learning experience to become strictly procedural.

A unit of instruction connects topics in mathematics that are naturally grouped together—the essential ideas or content standard clusters. The essential learning standards are framed as overarching questions for a unit posed to the class. It might take three to five days of instruction and two to three daily learning objectives to fully answer the essential question. The *context* of the lesson is the driving force for the entire lesson-design process. Each lesson context centers on clarity of the mathematical content and the processes for student learning.

The crux of any successful mathematics lesson rests on your collaborative team identifying and determining the daily learning objectives that align with the essential learning standards for the unit. Although you might develop daily learning objectives for each lesson as part of curriculum writing or review, your collaborative team should take time during lesson-design discussions to make sense of the essential learning standards for the unit and to consider how the essential learning standards for the unit are connected.

This involves unpacking the mathematics content as well as the Mathematical Practices or processes each student will engage in as he or she learns the mathematics of the unit. *Unpacking*, in this case, means making sense of the mathematics listed in the standard, making sense of how the content connects to content learned in other mathematics courses as well as within the current course, and making sense of how students might develop both conceptual understanding and procedural skill with the mathematics listed in the standard. Collaboratively unpacking the standards is one strategy to address one of the eight research-informed instructional practices identified by NCTM (2014) in *Principles to Actions: Ensuring Mathematical Success for All,* "Establish Mathematics Goals to Focus Learning." When collaborative teams unpack a standard and situate the "learning goals within the mathematical landscape" they support students in making mathematical connections and developing deep understanding of the content (p. 13).

This first high-leverage team action supports NCTM's curriculum principle and professionalism principle: *teachers collaboratively examine and prioritize the mathematics content and mathematical practices that students are to learn* (2014, p. 99). For more detail on these connections, see appendix E, page 189.

The How

As you and your collaborative team unpack the mathematics content standards (the essential learning standards) for a unit, it is also important to decide which Standards for Mathematical Practice will receive focused development throughout the unit of instruction and what mathematical tasks you will use during the unit to help students learn both the essential content standards and the Mathematical Practices or process standards. Thus, your collaborative team identifies, explores, and discusses:

1. The meaning of the essential *content* learning standards for the unit
2. The intentional Mathematical Practices or processes for student learning to be developed during the unit
3. The mathematical tasks (higher- and lower-level cognitive demand) to be used during the unit

Unpacking a Learning Standard

How can your team explore the general unpacking of content and linking the content to student Mathematical Practices for any unit? By participating in deep discussions about the meaning of the essential learning standards before the unit begins.

In order to develop clarity around the mathematical content and practices for any given unit, it is necessary to follow a process of unpacking the essential learning standards. As you organize each unit for the course, your team should identify learning objectives that develop understanding for the essential standards, Mathematical Practices and processes for learning, and specific instructional strategies you will commit to use as part of your daily instruction.

Making Sense of Essential Learning Standards for a Unit

As you read through the sample unit in figure 1.2 (page 12), you will notice that the essential learning standards offer clarity on the depth of instruction and focus of the content. In general, there is coherence among the essential learning standards as to how they progress through the grades 6–8 domains, such as

Ratios, Proportions, and Functions; Expressions and Equations; or Geometry. In order to deeply understand the intent within each essential learning standard of a unit, it is important to consider and reference the essential learning standards within that domain for the preceding and succeeding grade or course. Visit the University of Arizona's website (http://ime.math.arizona.edu/progressions) if you need more information on these progressions for the CCSS. Exploration of the mathematics standards at this grain size helps you and your team members develop confidence in exploring the mathematics you will teach and discussing uncertainties regarding the depth of conceptual understanding the standard requires.

Consider the eighth-grade mathematics unit aligned to the content standard cluster *Understand congruence and similarity using physical models, transparencies, or geometry software* (8.G.1–8.G.5; NGA & CCSSO, 2010, pp. 55–56) shown in figure 1.2.

Unit Name: Congruence and Similarity

Unit Number: 16

Essential Learning Standards

1. **Understand congruence and similarity using physical models, transparencies, or geometry software.**

    8.G.1: Verify experimentally the properties of rotations, reflections, and translations:
        a. Lines are taken to lines, and line segments to line segments of the same length.
        b. Angles are taken to angles of the same measure.
        c. Parallel lines are taken to parallel lines.

    Develop Mathematical Practices 5 and 6: "Use appropriate tools strategically" and "Attend to precision."

    8.G.2: Understand that a two-dimensional figure is congruent to another if the second can be obtained from the first by a sequence of rotations, reflections, and translations; given two congruent figures, describe a sequence that exhibits the congruence between them.

    (Prerequisite knowledge: 7.G.2)

    8.G.3: Describe the effect of dilations, translations, rotations, and reflections on two-dimensional figures using coordinates.

    Develop Mathematical Practices 3 and 7: "Construct viable arguments and critique the reasoning of others" and "Look for and make use of structure."

    8.G.4: Understand that a two-dimensional figure is similar to another if the second can be obtained from the first by a sequence of rotations, reflections, translations, and dilations; given two similar two-dimensional figures, describe a sequence that exhibits the similarity between them.

    Develop Mathematical Practices 3 and 6: "Construct viable arguments and critique the reasoning of others" and "Attend to precision."

    (Prerequisite knowledge: 7.G.1)

    8.G.5: Use informal arguments to establish facts about the angle sum and exterior angle of triangles, about the angles created when parallel lines are cut by a transversal, and the angle-angle criterion for similarity of triangles. *For example, arrange three copies of the same triangle so that the sum of the three angles appears to form a line, and give an argument in terms of transversals why this is so.*

    (Prerequisite knowledge: 7.G.5)

Source for standards: NGA & CCSSO, 2010, pp. 55–56.

Figure 1.2: Essential learning standards for a grade 8 geometry unit.

In this specific case for congruence and similarity, some team members may already have deep knowledge of this level of geometry, and others may only know about the essential learning standards on a very superficial level. For example, what is the intended congruence and similarity difference between 8.G.2 and 8.G.4? Some team members may not yet comprehend the depth of understanding they must have in order to meet this essential learning standard for their students.

You can use the congruence and similarity unit discussion tool in figure 1.3 to help inform your team understanding of the congruence and similarity standards. You can also examine an appropriate progression of essential learning standards and identify the depth of the essential learning standards for developing student knowledge around congruence and similarity of two-dimensional figures through dilations, rotations, translations, and reflections. Additionally, you and your collaborative team can plan and discuss *how* you want students to demonstrate understanding of the mathematics content through the identified Mathematical Practices or processes for the unit.

Directions: Work with your collaborative team to answer the following questions based on your team decisions regarding unpacking grade 8 geometry standards you will teach during a congruence and similarity unit from figure 1.2.

1. What do congruence and similarity mean? Do we have a common understanding of these concepts as a team?

2. How do we help students understand the difference between translation, rotation, reflection, and dilation?

3. What is the difference between demonstrating two shapes are congruent through a sequence of transformations versus demonstrating two shapes are similar through a sequence of transformations?

4. What instructional strategies can you use to help students transition to using a coordinate grid for describing transformations? What other grade 7 or 8 content domains or clusters naturally link with this unit?

5. Which Mathematical Practices or processes should we highlight during this unit? Do you agree with the choices indicated by this district and the grade 8 teacher team?

6. Do you need additional resources to gain clarity on any of these standards?

Figure 1.3: Discussion tool on essential learning standards for a congruence and similarity unit in grade 8.

Visit **go.solution-tree.com/mathematicsatwork** to download a reproducible version of this figure.

In order to help your collaborative team build lessons, it is helpful to break down the essential standards of a unit into the daily learning objectives, which in turn will help you build daily lessons using appropriate mathematical tasks that represent the standards.

The key element of this first high-leverage team action is to personally and collaboratively make sense of the essential learning standards with an eye toward *planning* for student engagement in the Mathematical Practices and processes that support them. This needs to occur *before* the unit begins in order to take full advantage of instructional time during the unit. In the case of the content standard cluster from domain 8.G, without your collaborative team's focus on unpacking the essential learning standards, students might be limited to demonstrating how a shape rotates, translates, or reflects. If this were to happen, students will not engage in the higher-level process of describing a sequence of movements for how one shape is transformed to a new location.

Your collaborative team may need to use outside resources to make sense of the mathematics involved in the learning standards within a unit. The background information in your school textbook or digital teacher's editions can be a good source for this foundational knowledge, as can resources from the National Council of Teachers of Mathematics (www.nctm.org), such as their *Essential Understanding* series.

In general, your team can use figure 1.4 as a discussion tool for any unit that is part of your grade level or course curriculum, as you break down the major essential learning standards for understanding.

After using the discussion tool in figure 1.4, you and your collaborative team can use the results of your conversations to create a transparent map of the unit and to articulate the unit intent to all team members. When unpacking the essential learning standards, your collaborative team develops understanding of the essential learning standards, more specific daily learning objectives, and the necessary prerequisite knowledge and vocabulary and identifies appropriate mathematical practices to support student learning.

Directions: Discuss the following prompts or questions with your collaborative teams to unpack essential learning standards, the prerequisite skills for the unit, the associated Mathematical Practices or processes relevant to the current unit of study for your course, and the pacing decisions for the unit.

1. List the agreed-on four to six essential learning standards for this unit.

2. As you discuss each essential learning standard, what are the daily learning objectives that might support that standard over several days?

3. What is the time frame available to teach this unit, and how will that time be distributed for each essential learning standard?

4. What prerequisite mathematics knowledge is necessary to support student learning during this unit?

5. What is the mathematics vocabulary necessary to support student learning during this unit?

6. What are specific teaching strategies, tasks, and tools that will most effectively support each essential learning standard for the unit?

7. Which Mathematical Practices or processes should we highlight during the unit in order to better engage students in the process of understanding each essential learning standard? Identify them, and discuss.

8. What specific lessons will highlight mathematical modeling that represents the standards for the unit?

Figure 1.4: Discussion tool for making sense of the agreed-on essential learning standards for the unit.

Visit **go.solution-tree.com/mathematicsatwork** to download a reproducible version of this figure.

Making Sense of the Unit Content Progression

Figure 1.5 is a sample unit plan designed to support HLTA 1—Making sense of the agreed-on essential learning standards—for a grade 8 unit that addresses congruence and similarity.

Unit Name: Congruence and Similarity	**Unit Number:** 1
Time Frame	
Twenty fifty-minute class periods (including the review and test)	
Essential Learning Standards	**Potential Learning Objectives for the Unit**
Understand congruence and similarity using physical models, transparencies, or geometry software.	

8.G.1: Verify experimentally the properties of rotations, reflections, and translations:
  a. Lines are taken to lines, and line segments to line segments of the same length.
  b. Angles are taken to angles of the same measure.
  c. Parallel lines are taken to parallel lines.
	I can verify experimentally the properties of rotations, reflections, and translations:
	  a. Lines are taken to lines, and line segments to line segments of the same length.
	  b. Angles are taken to angles of the same measure.
	  c. Parallel lines are taken to parallel lines.

Mathematical Practices Use appropriate tools strategically. Attend to precision.	I can use tools appropriately to model and investigate the properties of rotations, reflections, and translations. I can attend to precision in my use of proper vocabulary and properties.
8.G.2: Understand that a two-dimensional figure is congruent to another if the second can be obtained from the first by a sequence of rotations, reflections, and translations; given two congruent figures, describe a sequence that exhibits the congruence between them. (Prerequisite knowledge: 7.G.2)	I can demonstrate the congruence of two-dimensional figures using the properties of rotations, reflections, and translations.
8.G.3: Describe the effect of dilations, translations, rotations, and reflections on two-dimensional figures using coordinates.	I can describe the effect of translations, rotations, and reflections on two-dimensional figures using coordinates. I can describe the effect of dilations on two-dimensional figures using coordinates.
Mathematical Practices Construct viable arguments and critique the reasoning of others. Look for and make use of structure.	In order to identify the effect of transformations, I am able to look for and make use of structure. I can construct viable arguments and critique the reasoning of others to describe the effect of transformations on a given figure.
8.G.4: Understand that a two-dimensional figure is similar to another if the second can be obtained from the first by a sequence of rotations, reflections, translations, and dilations; given two similar two-dimensional figures, describe a sequence that exhibits the similarity between them. (Prerequisite knowledge: 7.G.1)	I can demonstrate that two figures are similar by using the properties of dilations, rotations, reflections, and translations of two-dimensional figures. I can describe a sequence of transformations between two figures that exhibits the similarity between them.
Mathematical Practices Construct viable arguments and critique the reasoning of others. Attend to precision.	I can construct viable arguments and critique the reasoning of others as they describe that a two-dimensional figure is similar or not similar to another figure. I can attend to precision as I describe the sequence of transformations to demonstrate similarity.

continued →

Overarching Unit Mathematical Practices	I can demonstrate perseverance in all problems I encounter.
Make sense of problems and persevere in solving them.	
	I can generate models to represent my thinking and use models to help develop understanding.
Model with mathematics.	

| **Prerequisite Knowledge** |
| List standards linked to content taught in the previous grade or course. |

7.G.1: Solve problems involving scale drawings of geometric figures, including computing actual lengths and areas from a scale drawing and reproducing a scale drawing at a different scale.

7.G.2: Draw (freehand, with ruler and protractor, and with technology) geometric shapes with given conditions. Focus on constructing triangles from three measures of angles or sides, noticing when the conditions determine a unique triangle, more than one triangle, or no triangle.

7.G.5: Use facts about supplementary, complementary, vertical, and adjacent angles in a multi-step problem to write and solve simple equations for an unknown angle in a figure.

6.NS.6c: Find and position integers and other rational numbers on a horizontal or vertical number line diagram; find and position pairs of integers and other rational numbers on a coordinate plane.

| **Mathematical Practices** |

Make sense of problems and persevere in solving them.

Construct viable arguments and critique the reasoning of others.

Model with mathematics.

Attend to precision.

Look for and make use of structure.

| **Key Mathematics Vocabulary** |

- Rotation
- Reflection
- Translation
- Dilation
- Congruent
- Similarity, similar

- Two-dimensional
- Coordinate plane
- Coordinates
- Angles, lines, line segments, parallel lines
- Transformations

Source: Adapted from D125 Consortium, Lincolnshire, Illinois.
Source for standards: NGA & CCSSO, 2010, pp. 43, 49–50, 55–56.

Figure 1.5: Sample geometry unit plan for congruence and similarity.

Once your team identifies the essential standards for the unit (notice that the team in the grade 8 example decided not to teach essential standard 8.G.5 in this unit due to length issues), you turn your focus to developing the essential learning standard progression of the unit. You decide how much time to dedicate to each essential learning standard. As with any planning, this is a starting point to understanding the depth of student understanding required for the unit and how to organize the development of the mathematical concepts within the unit. Your team shares how each concept is connected to previous standards and upcoming standards to make explicit and logical connections among the unit's content for the students. Figure 1.6 (pages 18–19) provides a sample unit progression across these standards, for use with grade 8 students.

Unit Plan: Twenty Instructional Days

Day 1	Day 2	Day 3	Day 4	Day 5
8.G.1:	**8.G.1:**	**8.G.1:**	**8.G.1:**	**8.G.2:**
I can verify experimentally the properties of rotations, reflections, and translations.	I can verify experimentally the properties of rotations, reflections, and translations.	I can verify experimentally the properties of rotations, reflections, and translations.	I can verify experimentally the properties of rotations, reflections, and translations.	I can demonstrate the congruence of two-dimensional figures using the properties of rotations, reflections, and translations.
Exploration Using Geometry Software	Exploration Using Geometry Software	Informal Assessment	Students will engage in a final vocabulary activity using the Frayer Model to finalize understanding about the properties of transformations.	Students will discuss congruence and begin creating congruent figures through translations.
Students will manipulate shapes through translations and reflections to make conjectures about their observations.	Students will manipulate shapes through rotations to make conjectures about their observations.	Students create a transformation to prove the properties of transformations.		

Day 6	Day 7	Day 8	Day 9	Day 10
8.G.2:	**8.G.2:**	**8.G.2:**	**8.G.2:**	**8.G.2:**
I can demonstrate the congruence of two-dimensional figures using the properties of rotations, reflections, and translations.	I can demonstrate the congruence of two-dimensional figures using the properties of rotations, reflections, and translations.	I can demonstrate the congruence of two-dimensional figures using the properties of rotations, reflections, and translations.	I can demonstrate the congruence of two-dimensional figures using the properties of rotations, reflections, and translations.	Informal Assessment
Students will explore congruent figures that are reflected over the x-axis and y-axis.	Students will explore congruent figures that are reflected over other lines.	Students will explore rotations of shapes and identify if they are congruent.	Students will create rotations that result in congruent figures and noncongruent figures using physical models and geometry software.	I can demonstrate the congruence of two-dimensional figures using the properties of rotations, reflections, and translations.
				Students will create a sequence of transformations for a peer to decide if they are congruent.

continued →

Day 11	Day 12	Day 13	Day 14	Day 15
8.G.3:	**8.G.3:**	**8.G.4:**	**8.G.4:**	**8.G.4:**
I can describe the effect of translations, rotations, and reflections on two-dimensional figures using coordinates. Students will apply their knowledge of transformations and learn how to use coordinates to describe a transformation or a series of transformations.	I can describe the effect of translations, rotations, and reflections on two-dimensional figures using coordinates. Students will apply their knowledge of transformations and learn how to use coordinates to describe a transformation or a series of transformations.	I can demonstrate that two figures are similar by using the properties of dilations, rotations, reflections, and translations of two-dimensional figures. Students will begin to explore the meaning of similar figures and the difference between similar and congruent.	I can demonstrate that two figures are similar by using the properties of ~~dilations,~~ rotations, reflections, and translations of two-dimensional figures. Students will discuss how two figures can be similar using reflections and translations. They will examine examples and nonexamples.	I can demonstrate that two figures are similar by using the properties of ~~dilations,~~ rotations, reflections, and translations of two-dimensional figures. Students will discuss how two figures can be similar using rotations, reflections, and translations. They will examine examples and nonexamples.

Day 16	Day 17	Day 18	Day 19	Day 20
8.G.3 (Part 2):	**8.G.3 (Part 2) and 8.G.4:**	**8.G.4:**	**Review for Unit 1**	**Assessment for Unit 1**
I can describe the effect of dilations on two-dimensional figures using coordinates. Students will explore the effect of dilations on coordinates for two-dimensional figures using models and geometry software. Students will establish generalizations about effects.	I can describe the effect of dilations on two-dimensional figures using coordinates. Students will continue building their knowledge of dilations and how to represent the effect of a dilation using coordinates through various tasks.	I can describe a sequence of transformations between two figures that exhibits the similarity between them. Students will be given and will create a sequence of transformations between two figures and describe the sequence.	Students will combine all standards together.	

Notes for Unit 1

When working through each standard, we may not need to break up the learning targets by each transformation; however, it may also help students to take an in-depth look at each transformation. This is something we will monitor throughout the unit and make notes on for next year. Also, 8.G.1 will continue to be embedded throughout instruction in this unit. Before moving on to similarity, we will ensure all students have a solid understanding of congruence and how it relates to transformations. Most work and dialogue during this unit will occur in teams of four. Students will present their thinking and listen to the thinking and reasoning of others to fully develop their understanding and their demonstration for the overarching unit, Mathematical Practices 1 "Make sense of problems and persevere in solving them" and 4 "Model with mathematics."

Note: Crossed-out text indicates that only a certain portion of the standard is the focus.

Source: Adapted with permission from Aptakisic-Tripp CCSD 102, Buffalo Grove, Illinois.

Source for standards: NGA & CCSSO, 2010, pp. 55–56.

Figure 1.6: Sample unit progression for geometry across standards.

Visit **go.solution-tree.com/mathematicsatwork** to download a reproducible version of this figure.

The first high-leverage team action is both a district and a teacher team responsibility. The district office does need to provide guidance about the proper scope and sequence of the essential learning standards of the unit to ensure students receive a guaranteed and viable curriculum and to support student mobility issues. At the same time, members of your course- or grade-level team need to be clear on the intent of the essential standards, the rationale for teaching the standards in a specific order, and the nuances of the meaning and intent of each essential standard. In short, you need to own the meaning of each essential learning standard for the unit.

At a minimum, your team should take the time to write a set of notes similar to the ones stated at the end of figure 1.6 as you work to better understand the intent of the mathematics content, the mathematics content progressions, the overarching Mathematical Practices and processes, and the mathematical tasks and applications for that unit.

Your Team's Progress

It is helpful to diagnose your team's current reality and action prior to launching the unit. Ask each member to individually assess your team on the first high-leverage team action using the status check tool in table 1.1. Discuss your perception of your team's progress on making sense of the agreed-on essential learning standards and pacing. It matters less which stage your team is at and more that you and your team members are committed to working together to focus on understanding the essential learning standards and the best mathematical tasks and strategies for increasing student understanding and achievement as your team seeks stage IV—sustaining.

Your responses to table 1.1 will help you determine what you and your team are doing well with respect to your focus on essential learning standards and where you might need to place more attention before the unit begins.

Once your team unpacks and understands the essential learning standards, you are ready to identify and prepare higher-level-cognitive-demand mathematical tasks related to those essential learning standards. It is necessary to include tasks at varying levels of demand during instruction. The idea is to match the tasks and their cognitive demand to the essential learning standard expectations for the unit. Selecting mathematical tasks together is the topic of the second high-leverage team action, HTLA 2.

Table 1.1: Before-the-Unit Status Check Tool—HLTA 1: Making Sense of the Agreed-On Essential Learning Standards (Content and Practices) and Pacing

Directions: Discuss your perception of your team's progress on the first high-leverage team action—making sense of the agreed-on essential learning standards (content and practices) and pacing. Defend your reasoning.			
Stage I: Pre-Initiating	**Stage II: Initiating**	**Stage III: Developing**	**Stage IV: Sustaining**
We do not discuss the essential learning standards of the unit prior to teaching it.	We discuss and reach agreement on the four to six essential learning standards for the unit.	We unpack the intent of each essential learning standard for the unit and discuss daily learning objectives to achieve each essential standard.	We connect the four to six essential learning standards to the Mathematical Practices before the unit begins.
We do not know which essential learning standards other colleagues of the same course or grade level teach during the unit.	We discuss and share how to develop student understanding of the essential learning standards during the unit.	We collaborate with our colleagues to make informed decisions about instruction of the essential learning standards for each lesson in the unit.	We have procedures in place to review the effectiveness of the students' roles, activities, experiences, and success on the essential learning standards during the unit.
We do not discuss lesson tasks.	We connect and align some lesson tasks to the essential learning standards for the unit.	We share effective teaching strategies for the essential learning standards of the unit.	We have procedures in place that ensure our team aligns the most effective mathematical tasks and instructional strategies to the content progression established in our overall unit plan components.
We do not discuss Mathematical Practices and processes as part of our unit planning.	We discuss Mathematical Practices and processes that best align to the essential learning standards for the unit.	We agree on Mathematical Practices and processes that best align to the learning standards for the unit.	We implement Mathematical Practices and processes that best align to the learning standards for the unit.

Visit **go.solution-tree.com/mathematicsatwork** to download a reproducible version of this table.

2: Identifying Higher-Level-Cognitive-Demand Mathematical Tasks

The function of education is to teach one to think intensively and to think critically.

—Martin Luther King Jr.

Developing your team's understanding of the essential learning standards for the unit helped you answer the first critical question of a PLC, What do we want all students to know and be able to do? The *mathematical tasks* you and your team choose to use every day during the unit help you answer this first critical question as well.

The mathematical tasks you choose each day for every unit also partially answer the second critical question of a PLC, How will we know if they know it?

High-Leverage Team Action	1. What do we want all students to know and be able to do?	2. How will we know if they know it?	3. How will we respond if they don't know it?	4. How will we respond if they do know it?
Before-the-Unit Action				
HLTA 2. Identifying higher-level-cognitive-demand mathematical tasks	▨	▧		

▨ = Fully addressed with high-leverage team action

▧ = Partially addressed with high-leverage team action

The What

What is a mathematical task?

NCTM first identified the term *mathematical task* in its (1991, 2008) *Professional Teaching Standards* as "worthwhile mathematical tasks" (p. 24). Melissa Boston and Peg Smith (2009) later provide this succinct definition: "A mathematical task is a single complex problem or a set of problems that focuses students' attention on a specific mathematical idea" (p. 136).

Mathematical tasks include activities, examples, or problems to complete as a whole class, in small groups, or individually. The tasks provide the rigor (levels of complex reasoning as provided by the conceptual understanding, procedural fluency, and application of the tasks) that students require and thus become an essential aspect of your team's collaboration and discussion. In short, the tasks are the problems you choose to determine the pathway of student learning and to assess student success along that pathway. As a teacher, you are empowered to decide what and how a student learns through your choice and use of the mathematical tasks and activities that students experience.

The type of instructional tasks you and your team select and use will determine students' opportunities to develop proficiency in Mathematical Practices and processes and will support the development of conceptual understanding and procedural fluency for the essential learning standards. Your provision of higher-level-cognitive-demand tasks in lessons, and your sequencing of those tasks to build conceptual knowledge and procedural fluency, are essential aspects of your middle school mathematics lesson planning and lesson implementation.

The implementation of instructional tasks that promote reasoning and sense-making is so critical to student learning of mathematics that it is listed as one of the eight research-informed teaching practices in *Principles to Actions* (NCTM, 2014).

As Glenda Lappan and Diane Briars (1995) state:

> There is no decision that teachers make that has a greater impact on students' opportunities to learn and on their perceptions about what mathematics is than the selection or creation of the tasks with which the teacher engages students in studying mathematics. (p. 139)

Mathematical Practice 1—"Make sense of problems and persevere in solving them"—establishes the expectation for regularly engaging your students in challenging, higher- and lower-level-cognitive-demand tasks essential for their development. A growing body of research links students' engagement in higher-level-cognitive-demand mathematical tasks to overall increases in mathematics achievement, not just in the ability to solve problems (Hattie, 2012; Resnick, 2006).

A key collaborative team decision is which tasks to use in a particular lesson to help students attain the daily learning objective. The nature of the tasks with which your students engage provides the common student learning experiences you can draw on to further student learning at various points throughout the unit. Selecting appropriate tasks provides your collaborative team with the opportunity for rich, engaging, and professional discussions regarding expectations about student performance for the unit.

Thus, four critical task questions for your grade-level or course-based collaborative team to consider include:

1. What nature of tasks should we use for each essential learning standard of the unit? Will the tasks focus on building student conceptual understanding, procedural fluency, or a combination? Will the tasks involve application of concepts and skills?

2. What are the depth, rigor, order of presentation, and ways of investigating that we should use to ensure students learn the essential learning standards?

3. How does our collaborative team choose the mathematical tasks that best represent each essential learning standard?

4. How does our team ensure the implementation of the tasks as a team in order to avoid wide variances in student learning across the grade level or course?

Conceptual understanding *and* procedural fluency are essential aspects for students to become mathematically proficient. In light of this, the tasks you choose to form the unit's lessons must include a balance of higher- and lower-level-cognitive-demand expectations for students. Your team will also need to decide which mathematical tasks to use for class instruction and which tasks to use for the various assessment instruments given to students during and at the end of a unit.

Higher-level-cognitive-demand lessons or tasks are those that provide "opportunities for students to explain, describe, justify, compare, or assess; to make decisions and choices; to plan and formulate questions; to exhibit creativity; and to work with more than one representation in a meaningful way" (Silver, 2010, p. 2). In contrast, lessons or tasks with only lower-level cognitive demand are "characterized as opportunities for students to demonstrate routine applications of known procedures or to work with a complex assembly of routine subtasks or non-mathematical activities" (Silver, 2010, p. 2).

However, selecting a task with higher-level cognitive demand does not ensure that students will engage in rigorous mathematical activity (Jackson, Garrison, Wilson, Gibbons, & Shahan, 2013). The cognitive demand of a mathematical task is often lowered (perhaps unintentionally) during the implementation phase of the lesson (Stein, Remillard, & Smith, 2007). During the planning phase, your team should discuss how you will respond when students urge you to lower the cognitive demand of the task during the lesson. Supporting productive struggle in learning mathematics is one of the eight research-informed mathematics teaching practices outlined in *Principles to Actions* (NCTM, 2014), and strategies to avoid cognitive decline during task implementation are discussed further in chapter 2 (page 83, HLTA 6).

Thus, your teacher team responds to several mathematical task questions before each unit begins:

1. How do we define and differentiate between higher-level-cognitive-demand *and* lower-level-cognitive-demand tasks for each essential standard of the unit?

2. How do we select common higher-level-cognitive-demand and lower-level-cognitive-demand tasks for each essential standard of the unit?

3. How do we create higher-level-cognitive-demand tasks from lower-level-cognitive-demand tasks for each essential standard of the unit?

4. How do we use and apply higher-level-cognitive-demand tasks for each essential standard during the unit?

5. How will we respond when students urge us to lower the cognitive demand of the task during the implementation phase of the lesson?

Visit **go.solution-tree.com/mathematicsatwork** to download these questions as a discussion tool.

The How

A critical step in selecting and planning a higher-level-cognitive-demand mathematical task is working the task before using it with students. Working the task provides insight into the extent to which it will engage students in the intended mathematics concepts, skills, and Mathematical Practices and how students might struggle. Working the task with your team provides information about possible solution strategies or pathways that students might demonstrate.

Defining Higher-Level- and Lower-Level-Cognitive-Demand Mathematical Tasks

You choose mathematical tasks for every lesson, every day. Take a moment to describe how you choose the daily tasks and examples that you use in class. Do you make those decisions by yourself, with your team members, before the unit begins, the night before you teach the lesson? Where do you locate and choose your mathematical tasks? From the textbook? Online? From your district resources?

And, most importantly, how would you describe the rigor of each task you chose for your students? Rigor is not whether a problem is considered hard. For example, "Solve $3(5x - 14) = 18$" might be a hard problem for some middle school students, but it is not rigorous. *Rigor of a mathematical task* is defined in this handbook as the level and the complexity of reasoning required by the student during the task (Kanold, Briars, & Fennell, 2012). Rigor is not about memorizing routine procedures.

There are several ways to label the demand or rigor of a task; however, for the purpose of this handbook, tasks are classified as either lower-level cognitive demand or higher-level cognitive demand as defined by Margaret Smith and Mary Kay Stein (1998) in their task-analysis guide and printed in full as appendix C (page 185). *Lower-level-cognitive-demand tasks* are typically focused on memorization or performing standard or rote procedures without attention to the properties that support those procedures (Smith & Stein, 2011).

Higher-level-cognitive-demand tasks are tasks for which students do not have a set of predetermined procedures to follow to reach resolution, or, if the tasks involve procedures, they require that students provide the justification for why and how the procedures can be performed. Smith and Stein (2011) describe these procedures as "procedures with connections" (p. 16) as opposed to "procedures without connections," the designation they use for lower-level-cognitive-demand tasks that are not just based on memorization.

Lower-level-cognitive-demand tasks generally take less time in class and do not require much complex reasoning by students. Their efficiency is appealing. They are much easier to manage in class as a general rule and easily serve direct instruction from the front of the room.

Consider the mathematical task presented in figure 1.7 (page 26), a sixth-grade problem. Examine the criteria listed in appendix C on page 185 and discuss with your colleagues why you believe the Alex problem in figure 1.7 is an example of a higher-level-cognitive-demand task. Which of the higher-level-cognitive-demand criteria seem to be met by this mathematical task for sixth graders? How is this task different from merely asking your students to do the lower-level-cognitive-demand task, "Find ¼ ÷ ⅖"?

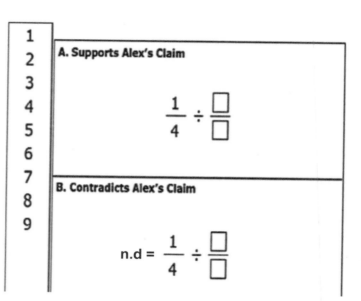

Source for the task: Smarter Balanced Assessment Consortium, n.d. Used with permission.

Figure 1.7: Sixth-grade example of a higher-level-cognitive-demand task.

Ultimately, the level of cognitive demand of the mathematical tasks you choose each day can be viewed as either lower- or higher-level-cognitive-demand as shown in figure 1.8.

> **Lower-Level Cognitive Demand**
>
> *Memorization:* Requires eliciting information such as a fact, definition, term, or a simple procedure, as well as performing a simple algorithm or applying a formula.
>
> *Procedures without connections:* Requires the engagement of some mental processing beyond a recall of information.
>
> **Higher-Level Cognitive Demand**
>
> *Procedures with connections:* Requires complex reasoning, planning, using evidence, and explanations of thinking.
>
> *Doing mathematics:* Requires complex reasoning, planning, developing, and thinking most likely over an extended period of time.

Source: Smith & Stein, 2012.

Figure 1.8: Four categories of cognitive demand.

Visit **go.solution-tree.com/mathematicsatwork** to download a reproducible version of this figure.

Since many of the revised state mathematics assessments intend to dramatically increase the task rigor compared to current state assessments (Herman & Linn, 2013) there are additional expectations for you to increase the cognitive demand of the mathematical tasks you choose to use during instruction and assessment.

The very nature of the mathematical content expectations requires your students to demonstrate *understanding*, and thus a shift to a *balanced* task approach during the unit—the use of both higher- and lower-level-cognitive-demand tasks. In most middle school classrooms, this will require an increase in the use of higher-level-cognitive-demand tasks.

Identifying the Cognitive Demand of Your Mathematical Tasks

As a first step in understanding the nature of the current cognitive-demand level of the tasks you use each day, use the tool in figure 1.9 for sorting unit tasks by cognitive-demand level.

Name of the Unit:	
For at least two of the essential standards in this unit, provide samples of the types of mathematical tasks students will experience in class, for homework, or on assessments.	
Directions: Sort every task you use into the following four categories.	

Lower-Level Tasks	**Higher-Level Tasks**
Memorization	**Procedures With Connections**
Procedures Without Connections	**Doing Mathematics**

Figure 1.9: Tool for sorting unit tasks by cognitive-demand level.

Visit **go.solution-tree.com/mathematicsatwork** to download a reproducible version of this figure.

What percent of the current tasks you plan to use fall into the lower-level-cognitive-demand task category? What percent fall into the higher-level-cognitive-demand task category? Do you have a proper balance in terms of the complexity of student reasoning required by the tasks you present to students throughout the unit?

Use figure 1.10 (page 28) to examine a seventh-grade equivalent expressions mathematical task, and then answer the questions at the end.

You can use your answers to the questions in figure 1.10 to guide your collaborative team's discussion about the use of any higher-level-cognitive-demand task during the unit.

Standard: I can understand the meaning of equivalent expressions.

Look at each expression. Is it equivalent to $\frac{x + 3y}{2}$?

Select Yes or No for expressions A–D.

A. $\frac{4x + 3y}{8}$ ○ Yes   ○ No

B. $\frac{5}{4}\left(\frac{2x + 6}{5}\right)$ ○ Yes   ○ No

C. $\frac{1}{2}(x + 3y)$ ○ Yes   ○ No

D. $\frac{2}{3}\left(\frac{5x}{6} + \frac{9y}{4} - \frac{x}{12}\right)$ ○ Yes   ○ No

Explain why each choice (A, B, C, and D) is equivalent or why it is not equivalent.

Directions: Find a solution pathway to the problem by yourself first, and then discuss the mathematics task with your collaborative team.

1. How are your collaborative team members' responses the same? How do they differ?

2. How does this task (and your solution pathway to the task) support the essential learning standard for equivalent expressions and what is the prerequisite knowledge needed for the task?

3. How does this task meet the criteria for higher-level cognitive demand?

4. Which Mathematical Practices or processes might students engage while solving this higher-level-cognitive-demand mathematical task?

5. Where might students get stuck when trying to work on this task together?

Source for the task: Smarter Balanced Assessment Consortium, n.d. Used with permission.

Figure 1.10: Sample higher-level-cognitive-demand-task discussion tool.

Visit **go.solution-tree.com/mathematicsatwork** to download a reproducible version of this figure.

Creating Higher-Level-Cognitive-Demand Tasks

Higher-level-cognitive-demand tasks are essential for improving student achievement. There are many resources online and in print that can provide examples of higher-level-cognitive-demand tasks for classroom use. For a list of these resources, see appendix D (page 187), or visit **go.solution-tree.com /mathematicsatwork** for a listing of these resources.

However, you and your team should also focus on creating tasks of varying cognitive demand as a team. This will not only ensure better alignment to the essential learning standards for the unit but will also empower your team to have greater ownership and understanding of task design and selection.

There are several strategies you can use to change a lower-level-cognitive-demand task to higher-level cognitive demand. You can use the strategies in figure 1.11 to adjust a mathematical task to a higher level of cognitive demand.

1. Use comparison questions. (When is one situation greater than, equal to, or less than another?)
2. Ask a question across multiple representations in a task.
3. Validate a solution pathway or approach.
4. Require students to provide justifications for (explain) their solutions.
5. Evaluate the error or reasoning in a student solution and provide a correct solution pathway.
6. Create a context. Ask students to write a word problem that creates a context for the given information.
7. Ask students to determine an expression to represent a situation.
8. Create an open-ended debate-type task, so that multiple student responses will satisfy a solution to the mathematical task.

Figure 1.11: Strategies for increasing the cognitive demand of tasks.

Visit **go.solution-tree.com/mathematicsatwork** to download a reproducible version of this figure.

The task from figure 1.10 for the essential learning standard 7.EE.1, "Apply properties of operations as strategies to add, subtract, factor, and expand linear expressions with rational coefficients" (NGA & CCSSO, 2010, p. 49), represents a higher-level-cognitive-demand task. The task as presented is an example of the first type of higher-level-cognitive-demand task listed in figure 1.11—use comparison questions.

Notice how the task question asks students if the expressions are equivalent by requiring students to compare the given expression to the options listed: a comparison of two quantities. What might a lower-level-cognitive-demand task look like for the same essential learning standard? Contrast the type of understanding required of the student when compared to the more typical lower-level-cognitive-demand mathematics question or task such as:

Simplify $\frac{1}{2}(2x + 3y)$ and show your work.

This is a lower-level-cognitive-demand mathematical task, as it only requires students to use the distributive property and represents a procedural question without a connection or expectation for understanding the procedure. In general, verbs such as *find*, *solve*, *graph*, and *simplify* tend to be lower-level-cognitive-demand verbs. *Create*, *build*, *conjecture*, *compare*, *contrast*, *explain*, and *justify* tend to create higher-level-cognitive-demand mathematical tasks.

Work with your course- or grade-level collaborative team to clarify or create higher-level-cognitive-demand tasks for the essential learning standards of your unit. You can use the team discussion tool in figure 1.12 to help with this process.

Once you include identifying and creating higher-level-cognitive-demand mathematical tasks as part of your unit planning, your before-the-unit activity will likely change. You will begin to look at tasks and classify them as higher-level or lower-level cognitive demand and then decide the best timing during the unit and lessons to use the higher-level-cognitive-demand tasks. Your goal should be to use an appropriate balance of higher- to lower-level-cognitive-demand questions during your instruction and in the assessment instruments for the unit in order to pursue "conceptual understanding, procedural skills and fluency, and application with equal intensity" (Common Core State Standards Initiative, 2014, p. 1). We will discuss the use of tasks in class during the unit in further detail in chapter 2, HLTA 6 (page 83).

Preparing for the Use of Higher-Level-Cognitive-Demand Tasks

Before you use any higher-level-cognitive-demand task in class, your teacher team should:

1. Discuss your expectations for *student demonstration of quality work* (both successful and unsuccessful approaches) in defense of their mathematical argument for the task.

2. Discuss how your lesson plan for this problem *promotes student communication of their argument with others* and allows peer-to-peer–based solution defense.

To help your team facilitate this type of discussion, you can use figure 1.13 (page 32) for any common higher-level-cognitive-demand task you plan to use during the unit.

To help your team facilitate this type of discussion, use the questions from figure 1.12 with the following higher-level-cognitive-demand task for grade 7 (see figure 1.14, page 33).

List an essential learning standard for the unit:

_____

Identify one lower-level-cognitive-demand task you use for this essential learning standard. Using one of the task-modification strategies from figure 1.10 (page 28), rewrite the task into a higher-level-cognitive-demand task. List both the lower- and higher-level-cognitive-demand task.

Provide a justification for your choice of tasks and the cognitive level of each task, and prepare to discuss these with your team.

Answer the following questions with your collaborative team:

1. How might what you learn about your students' understanding of the essential learning standard differ depending on the cognitive demand of the task you use during instruction?

2. Which strategy helped you to write the higher-level-cognitive-demand task?

3. In what ways might you support the implementation of the higher-level-cognitive-demand mathematical tasks during instruction? What types of teaching strategies or activities could you use?

Figure 1.12: Team discussion tool for identifying higher-level-cognitive-demand mathematical tasks for the unit.

Visit **go.solution-tree.com/mathematicsatwork** to download a reproducible version of this figure.

Directions: Use these questions to better understand how you will use any higher-level-cognitive-demand task in class.
What is the essential standard for the lesson? (What do you want students to know and understand about mathematics as a result of this lesson)?
In what ways does the task build on students' previous knowledge? What definitions, concepts, or ideas do students need to know to begin to work on this task? What prompts will you need to help students access their prior knowledge?
What are all the possible solution pathways for the task?   Which of these pathways or strategies do you think students will use?   What misconceptions might students have?   What errors might students make?
What are the language demands of the task? How will you address these challenges if students are stuck during the task?
What are your expectations for students as they work on and complete this task? What tools or technology will they utilize to enhance student-to-student discourse?

Source: Adapted from Smith, Bill, & Hughes, 2008.

Figure 1.13: Task-analysis discussion tool.

Visit **go.solution-tree.com/mathematicsatwork** to download a reproducible version of this figure.

Essential learning standard: I can solve problems involving scale drawings of geometric figures.

A floor plan is a scale diagram of a room or building drawn as if seen from above.

Part A

Task: On a separate piece of paper, create a floor plan of a house that has 2,500 square feet. There is a living room, kitchen, hallway, dining room, and a bathroom that make up this floor plan. Be sure to show all your work.

The floor plan must meet the following requirements:

The dining room is a perfect square.

The kitchen has dimensions of 35 ft. x 22 ft.

The bathroom has an area of 250 square feet.

The hallway is 3 ft. wide.

The living room is 3/10 of the total square feet.

Part B

If the volume of the living room is 7,875 cubic feet, how tall are the ceilings? Show all of your work.

Figure 1.14: Grade 7 higher-level-cognitive-demand task.

Visit **go.solution-tree.com/mathematicsatwork** to download a reproducible version of this figure.

Remember that what determines the cognitive demand of a task is the level and the complexity of reasoning required by the student during the task (Kanold, Briars, & Fennell, 2011). As you plan your lessons, consider the cognitive demand level a student is expected to reach when choosing your daily tasks. For example:

> • **Lower-level-cognitive-demand task—**
>
> Solve: $.4x + 0.03x = 1$
>
> • **Higher-level-cognitive-demand task—**
>
> In the equation $4x + 0.03x = 1$, a student attempts to eliminate the decimals by multiplying each term of the equation by 10. Is this an effective solution strategy? If yes, explain why. If not, explain why not and what you would recommend the student to do? Solve the problem correctly showing *all* work.

Since student work should be balanced with respect to the level of cognitive demand across tasks during every lesson, it is important to identify expected levels of cognitive demand and ultimately adapt or create tasks for each essential learning standard in the unit as the standards progress over time.

Your Team's Progress

It is helpful to diagnose your team's current reality and actions prior to launching the unit. Ask each team member to individually assess your team on the second high-leverage team action using the status check tool in table 1.2. Discuss your perception of your team's progress on identifying higher-level-cognitive-demand mathematical tasks. It matters less which stage your team is at and more that you and your team members are committed to working together to focus on understanding the learning standards and the best activities and strategies for increasing student understanding and achievement as your team seeks stage IV—sustaining.

Your responses will help your team focus on the cognitive demand for your daily mathematical tasks and where you need to place more time and attention before the unit begins. Your intentional use of higher-level-cognitive-demand mathematical tasks will ensure students are aware of and developing deeper understanding of the learning standards.

Of course, using balanced-cognitive-demand tasks becomes an important feature of the common assessment instruments for the end of the unit as well. Creating and using common assessment instruments with a balance of cognitive demand across tasks for each learning standard is the next high-leverage team action.

Visit nctm.org, see appendix D (p. 187), or visit **go.solution-tree.com/mathematicsatwork** for additional sources, resources, and examples of higher-level-cognitive-demand tasks for your grade level.

Table 1.2: Before-the-Unit Status Check Tool for HLTA 2—Identifying Higher-Level-Cognitive-Demand Mathematical Tasks

Directions: Discuss your perception of your team's progress on the second high-leverage team action—identifying higher-level-cognitive-demand mathematical tasks. Defend your reasoning.			
Stage I: Pre-Initiating	**Stage II: Initiating**	**Stage III: Developing**	**Stage IV: Sustaining**
We do not discuss or share our use of the mathematical tasks in each unit of the curriculum.	We discuss and share some mathematical tasks we will use during the unit.	We explore and practice together mathematical tasks we will use during the unit.	We reach agreement on a collection of mathematical tasks every team member will use.
We do not share our understanding of the difference between lower- and higher-level-cognitive-demand mathematical tasks.	We do not base our instructional decisions and mathematical task choices on the cognitive demand of the task.	We are able to compare and contrast higher- and lower-level-cognitive-demand mathematical tasks for each learning standard of the unit.	We reach agreement on both the solution pathways for each mathematical task and the management of those tasks in the classroom.
We do not discuss the cognitive demand of the tasks we use in class.	We have reached agreement on what differentiates a higher- from a lower-level-cognitive-demand mathematical task.	We connect the mathematical tasks to the essential learning standards, daily lesson learning objectives, and corresponding activities for each unit.	We choose mathematical tasks that represent a balance of lower- and higher-level cognitive demand for the learning standards of the unit.
We do not use higher-level-cognitive-demand mathematical tasks.	We use higher-level-cognitive-demand mathematical tasks if they are included in the lesson.	We create higher-level-cognitive-demand mathematical tasks from lower-level-cognitive-demand mathematical tasks individually.	We create higher-level-cognitive-demand mathematical tasks from lower-level-cognitive-demand mathematical tasks as a team.

Visit **go.solution-tree.com/mathematicsatwork** to download a reproducible version of this table.

HLTA 3: Developing Common Assessment Instruments

One of the most powerful, higher-level leverage strategies for improving student learning is the creation of frequent, higher-level quality, common formative assessments.

—Richard DuFour, Rebecca DuFour, Robert Eaker, and Tom Many

Just as the mathematical tasks you and your teacher team choose for your lessons help you to partially answer the second critical question of a PLC—How will we know if they know it?—so do the choices your team makes for the during-the-unit and end-of-unit common assessment instruments.

As your team makes sense of the essential learning standards for the unit and better understands how to choose, adapt, and create higher-level-cognitive-demand mathematical tasks and learning activities, your team will be ready to develop common assessment instruments to assess students' understanding of the essential learning standards for the unit.

High-Leverage Team Action	1. What do we want all students to know and be able to do?	2. How will we know if they know it?	3. How will we respond if they don't know it?	4. How will we respond if they do know it?
Before-the-Unit Action				
HLTA 4. Developing common assessment instruments	▨▢	▨		

▭ = Fully addressed with high-leverage team action

▨▢ = Partially addressed with high-leverage team action

The What

Why is developing common assessment instruments an important before-the-unit high-leverage activity? The process of creating common assessment instruments for each unit of your course supports your team conversations about prerequisite concepts and skills, common student errors, and ways of assessing students' understanding of the essential learning standards. It allows you to design lessons backward as you move from the outcomes (student demonstrations of knowledge on essential learning standards) for the unit to the learning activities, tasks, and resources students use during the unit and need for success on the end-of-unit assessments. Developing the common assessment instruments will also help you to better prepare for using them for a formative process at the end of the unit (discussed in detail in chapter 3 with HLTA 9, page 143) in order to inform feedback to students and guide instructional decisions in accord with NCTM's assessment principle outlined in *Principles to Actions* (NCTM, 2014).

According to DuFour, DuFour, Eaker, and Many (2010):

> One of the most powerful, high-leverage strategies for improving student learning available to schools is the creation of frequent, high-quality, common formative assessments by teachers who are working collaboratively to help a group of students acquire agreed-upon knowledge and skills. (p. 75)

There is an important distinction between formative assessment *processes* your team uses and the assessment *instruments* as part of those formative processes. W. James Popham (2011) provides an analogy to describe the difference between summative assessment instruments (such as your end-of-unit tests) and formative assessment processes (such as what you and your students *do* with those test results). He describes the difference between a surfboard and surfing.

> While a surfboard represents an important tool in surfing, it is only that—a part of the surfing process. The entire process involves the surfer paddling out to an appropriate offshore location, selecting the right wave, choosing the most propitious moment to catch the chosen wave, standing upright on the surfboard, and staying upright while a curling wave rumbles toward shore. (p. 36)

The surfboard is a key component of the surfing process, but it is not the entire process.

Your team's assessment instruments are the tools it uses to collect data about student demonstrations of the learning standards. The assessment instruments subsequently will inform you and your students' ongoing decisions about learning. Assessment instruments vary and can include such tools as class assignments, exit slips, quizzes, or unit tests; however, to avoid inequities in the level of rigor provided to students, and to serve the formative learning process, these assessment instruments must be *in common* for every teacher on your grade-level team.

When your collaborative team creates and adapts unit-by-unit common assessment instruments together, you enhance the coherence, focus, and fidelity to student learning expectations across all middle school courses. The common assessment instruments also provide coherence by fostering mathematics content learning progression continuity for students.

You minimize the wide variance in student task-performance expectations from teacher to teacher (an inequity creator) when you work collaboratively with colleagues to design high-quality assessment instruments appropriate to the identified essential learning standards for the unit. The expectation to collaborate on the development and use of common formative assessments is so critical as a support for equitable instruction that it is specifically listed as one of the actions mathematics teachers pursue in effective learning communities under the professionalism principle in NCTM's (2014) *Principles to Actions*.

The first questions your team must ask are, "How do we know our end-of-unit assessments are of high quality? On what basis would we make these determinations?"

The How

Collaborative teams consider the following when creating high-quality assessment instruments.

- What level of cognitive demand will we expect for each essential learning standard on the exam?

- What evidence of content knowledge will we assess for each essential learning standard?

- What evidence of student engagement in Mathematical Practices and processes will we assess for each essential learning standard?

- What types of question formats will we use to evaluate specific evidence of learning (such as multiple choice, short answer, multiple representations, explanation and justification, or technology)?

Once your team decides the types of questions or tasks you will use to understand student thinking, your team will need to develop high-quality common assessment instruments that reflect those decisions and support student use of the assessment instrument as a learning tool.

Evaluating the Quality of Your Current Assessment Instruments

How do you decide if the unit-by-unit assessment instruments you design are of high quality? Figure 1.15 provides a during- or end-of-unit assessment instrument quality-evaluation tool that your collaborative team can use to evaluate the quality of your current unit assessment instruments, such as tests and quizzes, as well as to build new and revised assessment instruments for each unit of the course.

Your collaborative team should rate and evaluate the quality of one of your most recent end-of-unit or chapter assessment instruments (tests) using the evaluation tool (figure 1.15) and the high-quality assessment diagnostic and discussion tool (figure 1.16, pages 40–41). How does it score—12? 16? 22? How close does your assessment instrument (your surfboard, so to speak) come to scoring a 27 or higher out of the thirty-two points possible in the rubric? It should be your expectation to write common assessment instruments that would score 4s in all eight categories of the assessment evaluation rubric.

Your collaborative team could also create agreed-on criteria for assessment instrument quality using figure 1.15 as a starting point, based on your local vision for high-quality assessment. Adapt the tool to fit your vision for high-quality assessments.

The value of any collaborative team–driven assessment depends on the extent to which the assessment instrument:

- Reflects the essential learning standards and clearly indicates those standards on the assessment instrument in student-friendly "I can . . ." language

- Supports a student formative process after the assessment (see HLTA 9 in chapter 3, page 143)

- Provides valid evidence of student learning for each essential standard

- Results in a positive impact on student motivation and continued learning

For additional practice using the assessment instrument quality-evaluation tool from figure 1.15, visit **go.solution-tree.com/mathematicsatwork** and see the end-of-unit sample assessments.

Designing a High-Quality Assessment Instrument

Designing common assessment instruments before the unit provides a context for the discussion of prerequisite knowledge, which you may need to address during instruction while making sense of the essential learning standards. It also provides a context for discussing potential student errors or misconceptions.

Your team can use the questions in figure 1.17 (page 42) as a way to unpack an essential learning standard and prepare for tasks that will need to be on your next common end-of-unit assessment instrument.

Assessment Indicators	Description of Level 1	Requirements of the Indicator Are Not Present	Limited Requirements of This Indicator Are Present	Substantially Meets the Requirements of the Indicator	Fully Achieves the Requirements of the Indicator	Description of Level 4
Identification and emphasis on essential learning standards (specific feedback to students)	Learning standards are unclear and absent from the assessment instrument. Too much attention is given to one target.	1	2	3	4	Learning standards are clear, included on the assessment, and connected to the assessment questions.
Visual presentation	Assessment instrument is sloppy, disorganized, difficult to read, and offers no room for work.	1	2	3	4	Assessment is neat, organized, easy to read, and well-spaced, with room for teacher feedback.
Balance of higher- and lower-level-cognitive-demand tasks	Emphasis is on procedural knowledge with minimal higher-level-cognitive-demand tasks for demonstration of understanding.	1	2	3	4	Test is rigor balanced with higher-level and lower-level-cognitive-demand tasks present.
Clarity of directions	Directions are missing and unclear. Directions are confusing for students.	1	2	3	4	Directions are appropriate and clear.
Variety of assessment task formats	Assessment contains only one type of questioning strategy, and no multiple choice or evidence of the Mathematical Practices. Calculator usage not clear.	1	2	3	4	Assessment includes a blend of assessment types and assesses Mathematical Practices modeling or use of tools. Calculator expectations clear.
Tasks and vocabulary (attending to precision)	Wording is vague or misleading. Vocabulary and precision of language are a struggle for student understanding and access.	1	2	3	4	Vocabulary is direct, fair, accessible, and clearly understood by students, and they are expected to attend to precision in response.
Time allotment	Few students can complete the assessment in the time allowed.	1	2	3	4	Test can be successfully completed in the time allowed.
Appropriate scoring rubric (points)	Scoring rubric is not evident or is inappropriate for the assessment tasks presented.	1	2	3	4	Scoring rubric is clearly stated and appropriate for each task or problem.

Source: Adapted from Kanold, Kanold, & Larson, 2012, p. 94.

Figure 1.15: Assessment instrument quality-evaluation tool.

Visit **go.solution-tree.com/mathematicsatwork** to download a reproducible version of this figure.

Directions: Examine your most recent end-of-unit test, and evaluate its quality against the following eight criteria described in figure 1.15 (page 39).

1. Are the essential learning standards written on the test as student friendly and grade-appropriate "I can . . ." statements?

Discuss: What do your students think about learning mathematics? Do your students think learning mathematics is about doing a bunch of math problems? Or, can they explain the essential learning standards and perform on any task that might reflect that standard?

Note: In order for students to respond to the end-of-unit assessment feedback when it is passed back (HLTA 9, in chapter 3), this is a necessary test feature.

2. Does the visual presentation provide space for student work?

Discuss: Do your students have plenty of space to write out solution pathways, show their work, and explain their thinking for each task on the assessment instrument?

Note: This criterion often is one of the reasons not to use the written tests that come with your textbook series. You can use questions from the test bank aligned to your instruction, but space problems as needed.

3. Is there an appropriate balance of higher- and lower-level-cognitive-demand questions on the test?

Discuss: What percentage of all tasks or problems on the assessment instrument are of lower-level cognitive demand? What percentage are of higher-level cognitive demand? Is there an appropriate balance? Unless this has been a major focus of your work, your current end-of-unit tests may not score very high in this criterion.

Note: Use figure 1.15 (page 39) as a tool to determine rigor. This will help you to better understand the level of cognitive demand. Also, see page 52 at the end of this section for more advice on this criterion. As a good rule of thumb, rigor-balance ratio should be about 30/70 (higher- to lower-level cognitive demand) on the assessment.

4. Is there clarity with all directions?

Discuss: What does clarity mean to each member of our team? Are any of the directions for the different test questions or tasks confusing to the student? Why?

Note: The verbs (actions words) used in the directions for each set of tasks or problems are very important to notice when discussing clarity.

5. Is there variety in assessment formats?

Discuss: Did our test use a blend of assessment formats or types? Did we include questions that allow for technology as a tool, such as graphing calculators? Did we balance the use of different question formats? If we used multiple choice, did we include items with multiple possible answers similar to those on the PARCC, SBAC, or other state assessments?

Note: Your end-of-unit assessments should not be of either extreme: all multiple-choice or all open-ended questions.

continued →

6. Is the language both precise and accessible?

Discuss: Is the vocabulary for each task used on our end-of-unit assessment clear, accessible, and direct for students? Do we attend to the precision of language used during the unit, and do the students understand the language used on the assessment?

Note: Does the assessment instrument place the proper language supports needed for all students?

7. Is enough time allotted for students to complete the assessment?

Discuss: Can our students complete this assessment in the time allowed? What will be our procedure if they cannot complete the assessment within the allotted time?

Note: Each teacher on the team should complete a full solution key for the assessment as will be expected of students. For upper-level students, it works well to use a time ratio of 3:1 (or 4:1) for student to teacher completion time to estimate how long it will take students to complete an assessment. For elementary students, it may take much longer to complete the assessment. All teachers should use the agreed-upon time allotment.

8. Are our scoring rubrics clear and appropriate?

Discuss: Are the scoring rubrics to be used for every task clearly stated on the test? Do our scoring rubrics (total points for the test) make sense based on the complexity of reasoning for the task? Are the scoring points assigned to each task appropriate and agreed upon by each teacher on the team?

Note: See HLTA 4 (page 54) for more details.

Summary: Using your score from the figure 1.15 assessment tool (page 39), which specific aspects of your current unit assessment instruments need to be improved?

_____

_____

_____

_____

_____

_____

_____

_____

_____

_____

_____

_____

_____

_____

Figure 1.16: High-quality assessment diagnostic and discussion tool.

Visit **go.solution-tree.com/mathematicsatwork** to download a reproducible version of this figure.

Directions: Choose an essential learning standard you are planning to assess in your next end-of-unit assessment, and answer the following questions. Be sure to look very carefully at the verbs that describe the essential standard. They will provide hints about the question or task types you will need for the test.

1. What prerequisite skills are necessary for this essential learning standard? How will you assess students' knowledge of these prerequisites?

2. What are common errors related to this essential learning standard? How will your instruction identify and resolve these errors before students take the common unit assessment?

3. How does your conversation around planning common assessment instruments influence your plans for instruction during the unit?

4. What mathematical tasks will you use during instruction for this essential learning standard, and what tasks will you reserve for the assessment of this standard?

Figure 1.17: Tool for planning and preparing for common assessment instrument task development.

Visit **go.solution-tree.com/mathematicsatwork** to download a reproducible version of this figure.

Once you identify and explore prerequisites and common errors, your team is better prepared to find or develop the common assessment instrument tasks or questions. If this is a new activity for your collaborative team, it might make sense to start with an existing assessment instrument and then adapt it so that it addresses the learning standards comprehensively and provides a balance of higher- and lower-level-cognitive-demand tasks.

Perhaps the greatest challenge you face with creating high-quality unit exams is finding the right balance between procedural fluency (mostly lower-level cognitive demand) with student demonstrations of understanding (mostly higher-level-cognitive-demand tasks) on each unit exam. Cognitive-demand balance is the third criterion listed in figure 1.15 (page 39), the assessment instrument quality-evaluation tool. Did the end-of-unit assessment instrument you reviewed using figure 1.15 score well in terms of an expected balance for student reasoning? Was there a blend of higher- and lower-level-cognitive-demand tasks and questions?

Figure 1.18 (pages 44–45) shows first-attempt sample assessment questions for an end-of-unit assessment instrument being developed for the grade 7 content standard clusters (7.EE) from a unit on expressions: *Use properties of operations to generate equivalent expressions* and *Solve real-life and mathematical problems using numerical and algebraic expressions and equations* (NGA & CCSSO, 2010, p. 49).

The specific essential learning standards for this end-of-unit assessment are:

1. Apply properties of operations as strategies to add, subtract, factor, and expand linear expressions with rational coefficients. (7.EE.1)

   ○ *Learning objective:* "I can apply properties of operations as strategies to factor and expand linear expressions with rational coefficients to generate equivalent expressions."

2. Understand that rewriting an expression in different forms in a problem context can shed light on the problem and how the quantities in it are related. (7.EE.2)

   ○ *Learning objective:* "I can rewrite expressions in different forms to show how quantities are related."

3. Solve multistep real-life and mathematical problems posed with positive and negative rational numbers in any form (whole numbers, fractions, and decimals), using tools strategically. Apply properties of operations to calculate with numbers in any form; convert between forms as appropriate; and assess the reasonableness of answers using mental computation and estimation strategies. (7.EE.3)

   ○ *Learning objective:* "I can solve real-life and mathematical problems using operations with rational numbers in any form."

4. Use variables to represent quantities in a real-world or mathematical problem, and construct simple equations and inequalities to solve problems by reasoning about the quantities. (7.EE.4)

   ○ *Learning objective:* "I can use variables to represent quantities in a real-world or mathematical problem, and construct simple equations to solve problems by reasoning about the quantities."

As you examine this assessment instrument, you will notice that it is *not* balanced with regard to cognitive demand or assessment-task formats, which are two criteria on the quality-evaluation assessment instrument tool. It also lacks some of the other specific criteria of a high-quality assessment. But for now, focus your team's discussion on the rigor-balance issue and the variety of task formats.

As you review this older version of an end-of-unit test, you will observe that students are being assessed on simplifying expressions using the distributive property and order of operations with integers. However, *simplify* is not part of the vocabulary within these standards. The intent of the seventh-grade standards 7.EE.1–4 is that students learn to *construct and use equivalent expressions* in both factored and expanded forms. This reality of the standards changes the nature of how you ask students questions on the assessment.

Name: _____ **Date:** _____ **Period:** _____

Grade 7 Unit 2 Sample Assessment

Noncalculator

Use the distributive property to simplify the expressions.

1. $4[10 - (1 + 7)]$　　　　2. $-2/5\ (x + 25)$　　　　　　3. $11(s + 9)$

4. $-21(x - 7)$　　　　5. $24y - 6(8 - 4y) + 52$　　　　6. $(4m + 9) - 3(2m - 5)$

Simplify each expression.

7. $6y + (-13y)$　　　　8. Subtract x from $3x - 1$.　　　　9. $4d - 5 - 9d + 17$

10. $27 - 13x + 32 - 2x + 10x$　　11. $-4(5x + 7) - 3x + 13$　　12. $7x^2 + 7y + 4x^2 - 4y$

Evaluate the expression when $x = 2$.

13. $15 + 3x + 10 + 8x$

14. Kirsten and her friends are going to the movies. Each person buys a ticket for $8, a medium drink for $2.75, and a large popcorn for $4.25.

　　a. Write an expression in simplest form that represents the amount of money each person spends at the movies. Use x to represent the total amount of people in the group.

continued →

15. Write and simplify an expression for the perimeter of the figure.

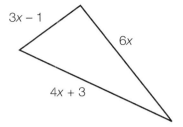

$3x - 1$

$6x$

$4x + 3$

Solve each equation.

16. $x + 5 = -7$

17. $-10 = z - 12$

18. $0 + (-21) = b$

19. $f + (-8) = 6$

20. $a + 5 + 8 = 20$

21. $-4n = -8$

22. $\frac{1}{3}x = 6$

23. $16 = -2x$

24. $3(x - 2) = -12$

25. $13.49 = -8.56 + y$

26. $30.2b = -75.5$

27. $\frac{x}{-2.1} = -7.5$

Choose the letter of the term that best matches each statement.

a. Terms

b. Coefficient

c. Constant term

d. Like terms

e. Equation

_____ 28. Terms that have identical variable parts

_____ 29. The parts of an expression that are added together

_____ 30. This type of term has a number but no variable.

_____ 31. The number part of the term

Figure 1.18: Sample assessment questions for end-of-unit test on expressions (lower-level cognitive demand only).

Visit **go.solution-tree.com/mathematicsatwork** to download a reproducible version of this figure.

Students are also being assessed on solving simple equations. A common student error when distributing and simplifying is that students often get confused when using positive and negative numbers, so notice the presence and location of integers. Also, the last few tasks on this test address some of the key vocabulary and language issues for this unit.

Referencing the essential learning standards, you will notice there are tasks in this sample test using rational numbers in both expressions and equations (see 7.EE.1 and 7.EE.3, page 43), along with a real-life expressions example. However, since students are not required to explain their reasoning, what is not assessed is student thinking and understanding with respect to the essential standards of the unit. There are also no mathematical tasks on the test that expect students to explain the process or solution pathway used to arrive at an answer or identify common errors. On another note, there are no real-life models of simple equations as expected in the essential learning standard for the unit.

Use figure 1.19 to work with your collaborative team to rewrite some of the items on the end-of-unit assessment instrument shown in figure 1.18 (pages 44–45). Use the questions from figure 1.17 (page 42) and the checking for cognitive-demand balance tool in figure 1.19, as a general team discussion guide for creating an improved balance between higher- and lower-level-cognitive-demand tasks and student opportunities to identify common errors related to the essential learning standards.

Directions: With your collaborative team, answer the following questions to check the cognitive-demand balance of your common assessment instruments.

1. What does the current assessment instrument do well in terms of the nature of the cognitive demand of each mathematical task on the test?

2. How are prerequisite skills and common misconceptions regarding the learning standards addressed in this assessment instrument?

3. Which tasks on the assessment should remain lower-level-cognitive-demand tasks?

4. Which tasks are more easily adapted into higher-level-cognitive-demand tasks? And how might you adapt them toward a higher level of cognitive demand?

Figure 1.19: Checking for cognitive-demand balance of common unit assessment instruments tool.

Visit **go.solution-tree.com/mathematicsatwork** to download a reproducible version of this figure.

The revised end-of-unit assessment in figure 1.20 (pages 47–51) for the seventh-grade content standard clusters *Use properties of operations to generate equivalent expressions* and *Solve real-life and mathematical problems using numerical and algebraic expressions and equations* (7.EE) provides an example of assessment questions similar to those questions (tasks) in figure 1.18 (pages 44–45), but it provides more balance with respect to cognitive demand and question format.

Notice that, for the most part, the problems are similar. However, students are asked to explain their reasoning and to find other ways to demonstrate their thinking. When asked to use a procedure, there is an expectation that students will connect the procedure to a demonstration of reasoning.

How does the revised end-of-unit assessment in figure 1.20 compare to the adjustments you made with your collaborative team? Your version does not need to match this assessment. What is important is that the assessment measures student performance on the learning standards, identifies potential misconceptions, provides students with an opportunity to demonstrate their depth of understanding, and is appropriately balanced for cognitive demand.

Name: _____ **Date:** _____ **Period:** _____

Grade 7: Unit 2 Assessment

Time: Fifty minutes

Tools allowed: Pencil, no calculator

Essential learning standard: I can use properties of operations to generate equivalent expressions (1–7).

- **7.EE.1**—I can apply properties of operations as strategies to factor and expand linear expressions with rational coefficients to generate equivalent expressions.
- **7.EE.2**—I can rewrite expressions in different forms to show how quantities are related.

Write an equivalent expression.

1. $11(s + 9)$

2. $-\frac{2}{5}(x + 25)$

3. Which of the following expressions is equivalent to $24y - 6(8 - 4y) + 52$? Show all your work and justify your reasoning.

   a. $28y + 4$

   b. $48y + 28$

   c. $0y + 4$ or 4

   d. $20y - 4$

   e. $48y + 4$

Figure 1.20: Revised sample assessment questions for end-of-unit test on expressions. continued →

4. Are the expressions $8x^2 + 3(x^2 + y)$ and $7x^2 + 7y + 4x^2 - 4y$ equivalent? Explain how you know.

5. A student solved the following problem incorrectly. Circle the mistake, explain what the student did wrong, and then solve the problem correctly. Remember to complete all the steps for full credit.

$(4m + 9) - 3(2m - 5)$

$= 4m + 9 - 6m - 15$

$= 4m - 6m + 9 - 15$

$= -2m - 6$

6. In the expression $-\frac{1}{4}x + 3$,

Gianna factored the expression and wrote: $-(\frac{1}{4}x - 3)$

Hannah factored the expression and wrote: $-1(\frac{1}{4}x - 3)$

Both Gianna and Hannah claim to be correct. Do you agree? Why or why not?

7. Which expression is not equivalent? Explain why.

a. Subtract x from $3x - 1$.

b. x more than $3x - 1$

c. $3x - 1$ decreased by x

d. Difference between $3x - 1$ and x

continued →

Essential learning standard: I can solve real-life mathematical problems using numerical and algebraic expressions and equations.

7.EE.3—I can solve real-life and mathematical problems using operations with rational numbers in any form.

7.EE.4—I can use variables to represent quantities in a real-world or mathematical problem, and construct simple equations to solve problems by reasoning about the quantities.

8. Photographers Ryan and Alex get paid per photography session. Ryan is paid a set-up fee of $25 plus $10 per hour. Alex is paid a set-up fee of $20 plus $11 per hour.

   Write an expression for each photographer. Use _h_ to represent hours worked.

Ryan: _____ Alex: _____

   a. Mr. Smith wants to have photos taken of his new baby boy. If the photo session will take 2 hours, who should he choose to take pictures? Explain your reasoning.

   b. If Mr. Smith wants to take pictures for 5 hours, whom should he choose as his photographer, Ryan or Alex? Explain your reasoning.

   c. If the photo session needed to go beyond 5 hours, would you change your answer from Part B? Explain why.

9. Kirsten and her friends are going to the movies. Each person buys a ticket for $8, a medium drink for $2.75, and a large popcorn for $4.25.

   a. Write an expression in simplest form that represents the amount of money each person spends at the movies. Use _x_ to represent the total amount of people in the group.

   b. If the total money spent was $60, how many people went to the movies?

continued →

10. Use the picture to answer the following questions.

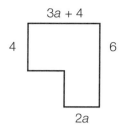

a. Write an expression, in simplest form, for the perimeter of the rectangle.

11. Javier's family decides to open a pizza place! The following chart shows the prices for a small cheese pizza plus additional toppings.

Toppings	Cost
Cheese	$12.00
Extra cheese	$1.00
Meat	$0.75
Veggies	$0.50

a. Write an expression to represent the cost of a cheese pizza with *v* for the veggie toppings. Then, identify how much a customer would spend on a veggie pizza.

b. Is it possible for someone to order a pizza for exactly $15? Explain your answer.

continued →

12. Madison wants to earn $350 for a new iPad mini. She already has $175 saved and has come up with a plan to earn the remaining amount.

   a. How much more money does Madison need to buy her iPad mini? Explain how you figured this out.

   b. Madison tells her parents that she will feed the dog two times each day for $1 each time. She also tells them she will walk the dog every day after school for $3. Write an expression to represent how much Madison will make each day.

   c. How much money will Madison make if she feeds her dog two times each day and walks the dog every day of the school week (Monday through Friday)? Show your work.

   d. How many school weeks must Madison do this to be able to earn the remaining amount? Show your work and explain how you arrived at your answer.

   e. Madison's mom offers Madison a new plan: if Madison walks the dog every day after school and feeds the dog two times each day from now until winter break (5 weeks), she will give Madison $200. Is this a better deal than what Madison offered? Explain your reasoning.

Solve each equation. Show all your work.

13. $x + 5 = -7$ 14. $\frac{1}{3}x = 6$ 15. $3(x - 2) = -12$

Source: Adapted with permission from Aptakisic Junior High School, Buffalo Grove, Illinois.
Source for standards: NGA & CCSSO, 2010, p. 49.

Figure 1.20: Revised sample assessment questions for end-of-unit test on expressions (continued).

You do not need to design your assessments from scratch; you can use assessment instruments provided with your curriculum materials and adjust them to ensure that they identify the essential learning standards, uncover common misconceptions, are appropriately balanced with respect to cognitive demand, provide the necessary time and space, and use appropriate, clear language and vocabulary.

Your collaborative team could also create your own agreed-on criteria for assessment instrument quality using figure 1.18 (pages 44–45) as a starting point based on your team or district's vision for higher-quality assessment. Use the tool in this handbook as a starting point, and adapt it to fit your vision for high-quality assessment instruments used in your mathematics program.

Your team will also need to consider the role of technology in your unit assessments. Since many state assessments are now administered in an online environment, it will be helpful if your students have some experience using technology as a tool during testing. Be sure to allow your students practice in the online format if those are expectations of your state assessment.

You should always look for resources that balance conceptual understanding, procedural fluency, and higher- and lower-level cognitive demand. Be sure that all materials you select support the mathematical understanding necessary to achieve the essential learning standards for the unit.

As resources to dig deeper into this issue, consult your state board of education website, NCTM (www.nctm.org), or the College Board (www.collegeboard.com/testing), or use sample online tests at Smarter Balanced Assessment Consortium (www.smarterbalanced.org), Partnership for Assessment of Readiness for College and Careers (PARCC) (www.parrconline.org), or the American College Testing Service (www.actaspire.org).

Your Team's Progress

As you and your collaborative team focus on developing common assessment instruments, remember that you do not need to design your assessment instruments from scratch. You can use the instruments provided with your curriculum materials and adjust them to ensure they address and list the essential learning standards, uncover common misconceptions, balance cognitive demand, can be completed in the available time, integrate technology as appropriate, and use appropriate and clear vocabulary.

It is helpful to diagnose your team's current reality and action prior to launching the unit. Ask each team member to individually assess your team on the third high-leverage team action using the status check tool in table 1.3. Discuss your perception of your team's progress on developing common assessment instruments. It matters less which stage your team is at and more that you and your team members are committed to working together to focus on understanding the learning standards and the best activities and strategies for increasing student understanding and achievement as your team seeks stage IV—sustaining.

Once you have prepared your common unit assessment, your team efforts should turn to creating a scoring rubric for the test and developing expected proficiency expectations for students. Developing scoring rubrics and proficiency expectations for the common assessment instruments is the fourth high-leverage team action in the before-the-unit-begins planning process. The process of developing scoring rubrics requires your team to reflect and stay focused on the essential learning standards for the unit.

Table 1.3: Before-the-Unit-Begins Status Check Tool for HLTA 3—Developing Common Assessment Instruments

Directions: Discuss your perception of your team's progress on the third high-leverage team action—developing common assessment instruments. Defend your reasoning.			
Stage I: Pre-Initiating	**Stage II: Initiating**	**Stage III: Developing**	**Stage IV: Sustaining**
We do not develop or use common assessment instruments.	Some members of our team develop common assessment instruments.	We develop common assessment instruments as a team, but not before the unit begins.	We design and write common assessments as a team before the unit begins.
We do not know if the end-of-unit assessments given by each member of the team are balanced for cognitive demand, provide sufficient time, and use clear language and vocabulary.	We develop end-of-unit common assessments connected to the learning standards, but they are not checked for balance of cognitive demand or clarity.	We develop common end-of-unit assessment instruments connected to the learning standards. They are either balanced for cognitive demand or clear but not both.	We develop common end-of-unit assessments that are clear, balanced, and connected to all aspects of the learning standards for the essential unit.
We do not know if our assessments are aligned to our instructional practices and reflect the essential learning standards of the unit.	We develop common assessments as a team, but not all members use them to influence their instructional plans for the unit.	Our planning for common assessments influences our instructional plans for the unit.	Our common assessments are deeply aligned with our instructional discussions and practices.

Visit **go.solution-tree.com/mathematicsatwork** to download a reproducible version of this table.

HLTA 4: Developing Scoring Rubrics and Proficiency Expectations for the Common Assessment Instruments

Do you trust me enough to allow me to grade your end-of-unit assessments?

—Tim Kanold

Creating a team culture of collaborative scoring and assessment discussions is one of the most important tasks of your middle school grade-level or course-based team. It ensures a greater chance for fidelity and accuracy in scoring all assessment instruments, and it eliminates the potential inequity a wide scoring variance from teacher to teacher can cause.

Just as the mathematical tasks and common assessment instruments (tests and quizzes) help you partially answer the second critical question of a PLC—How will we know if they know it?—so do the choices your team makes for scoring the mathematical tasks on the common unit assessments. The purpose of this team action will be discussed further in chapter 3, HLTA 9 and 10 (see page 141).

High-Leverage Team Action	1. What do we want all students to know and be able to do?	2. How will we know if they know it?	3. How will we respond if they don't know it?	4. How will we respond if they do know it?
Before-the-Unit Action				
HLTA 4. Developing scoring rubrics and proficiency expectations for the common assessment instruments		▣		

▣ = Partially addressed with high-leverage team action

The What

Why is this an important before-the-unit-begins high-leverage team activity? HLTA 4 will improve insight into the way you provide instruction pathways for students during the unit. More importantly, it will support valid inferences that you and your team make every day about students' knowledge—especially as you dedicate more time to higher-level-cognitive-demand tasks, and the complex reasoning required by your students. It will also improve the accuracy of your feedback and grading practices at the end of the unit and helps create greater team equity in the interpretation of student scores.

By reaching team agreement on the rubric score for each item on the end-of-unit test, you increase the reliability that the feedback for proficiency on the essential learning standards for the unit is accurate, and you increase your ability to ensure students understand the expectations of a solution pathway required to receive full credit on each task. Without this agreement, you and your team members are not feeding accurate data back into the system that you can use to provide feedback to students, guide effective instructional decisions, and make program improvements—the essence of the assessment principle in *Principles to Actions* (NCTM, 2014).

More important, this team action becomes an *inequity eraser* for your students and increases the likelihood that your feedback on their performance will be consistent and accurate across all members of your team and that it can be leveraged to improve student learning.

Determining how to score the assessment instrument involves far more than linking point values to test questions and tasks. As you work on scoring rubrics for tasks in your collaborative team, your instruction during the unit will benefit from:

- Discussing the value of each task relative to the other tasks on the test

- Deciding how you will determine if students have provided a complete solution for full credit relative to the essential learning standard each assessment task (problem) represents

- Deciding what you will do when students' answers are incomplete or incorrect—how will their response be scored?

These decisions are typically easier to make and more straightforward with lower-level-cognitive-demand tasks (as may have been the case with your past assessment instruments) but not as clear for the higher-level-cognitive-demand tasks necessary to measure student understanding, reasoning, *and* procedural fluency.

The How

A first step for your team is to examine a potential mathematical task (test question) and decide:

1. How many points to assign to the assessment task (the test question) to receive full credit

2. What level of student work or solution pathway would be required for the student to receive full credit

3. How to grade potential student solutions together in order to calibrate decisions for assigning points to the task

Closely examine the sample tasks shown in figures 1.21 and 1.22 (page 56). First, find a solution pathway that you would expect from a student in order to receive full credit. Assign a total number of points to the task, and be ready to explain your decisions. Then, calibrate your scoring rubric choice for the task by discussing solution pathways with your colleagues.

Directions: Find a solution pathway and decide how many points you would assign to this mathematical task for a student to receive full credit on the test. Justify your reasoning.

Jamal is filling bags with sand. All of the bags are the same size. Each bag must weigh less than 50 pounds. One sand bag weighs 57 pounds, and another sand bag weighs 41 pounds. Explain whether Jamal can put sand from one bag into the other so that the weight of each bag is less than 50 pounds.

Source: Reprinted from Smarter Balanced Assessment Consortium, n.d. Used with permission.

Figure 1.21: Sample grade 6 task.

Directions: Find a solution pathway and decide how many points you would assign to this mathematical task for a student to receive full credit on the test. Justify your reasoning.

Mr. Ruiz is starting a marching band at this school. He first does research and finds the following data about other local marching bands.

	Band 1	Band 2	Band 3
Number of Brass Instrument Players	123	42	150
Number of Percussion Instrument Players	41	14	50

Part A

Type your answer in the box. Backspace to erase.

Mr. Ruiz realizes that there are ⬚ brass instrument player(s) per percussion player.

Part B

Mr. Ruiz has 210 students who are interested in joining the marching band. He decides to have 80% of the band be made up of percussion and brass instruments. Use the unit rate you found in Part A to determine how many students should play brass instruments.

Show or explain all your steps.

Source: Partnership for Assessment of Readiness for College and Careers (PARCC), 2013.

Figure 1.22: Sample grade 6 task.

Use figure 1.24 (page 58) to work with your collaborative team to examine the photography task shown in figure 1.23.

7. Photographers Ryan and Alex get paid per photography session. Ryan is paid a set-up fee of $25 plus $10 per hour. Alex is paid a set-up fee of $20 plus $11 per hour.

a. Write an expression for each photographer. Use *h* to represent hours worked.

Ryan: _$25 + $10h_   Alex: _$20 + $11(h)_

b. Mr. Herrera wants to take pictures of his new baby boy. If the photo session will take 2 hours, who should he choose to take pictures? Explain your reasoning.

25 + 10(2)   20 + 11(2)   =   (Alex because it costs less money.
$45 $42   The flat fee is less and the per
                                     hour is only a dollar more.)

c. If Mr. Herrera wants to take pictures for 5 hours, should he choose Ryan or Alex as his photographer? Explain your reasoning.

25 + 10(5)   20 + 11(5)
$75 $75

   25   11
 + 50   x 5
 ———   ———
   75   Either photographer   55
             because both have the   + 20
             same price for 5 hrs.   ———
                                         75

Source: Adapted with permission from Aptakisic Junior High School, Buffalo Grove, Illinois.

Figure 1.23: Sample student response for photography task.

Now consider the photography task and the student sample work shown in the photography task in figure 1.23. How would you score, based on the complexity of reasoning required by the student, this mathematics task if it was on your test? Is it worth two points? Four points? Six points? Once that is decided, then how many points could the student receive for his or her response: 3 out of 4? 4 out of 6? 5 out of 6? Depending on your response, the student will receive a very different grade on this assessment task. Remember, accuracy of scoring is the goal. Begin by making sure your team agrees on the scoring value of the task, and then discuss the nature of the work students are expected to show to receive full credit for the task.

Directions: With your collaborative team, examine the following questions for the photography task shown in figure 1.23 (page 57).

1. How would you assign the points for different parts of the solution?

2. If the point value for the task is greater than one, how could a student get partial credit?

3. What evidence of student learning would receive full credit?

4. How will the team ensure the scoring of the task will be consistent between all team members?

5. How is the expected scoring rubric for the assessment task articulated to students?

6. Now score the student sample to determine consistency, and then share your score.

Figure 1.24: Collaborative team task scoring discussion prompts.

Visit **go.solution-tree.com/mathematicsatwork** to download a reproducible version of this figure.

How did your team determine scoring? Did students receive one point for their reasoning and one point for the correct answer, or no credit if the answer is incorrect? Is each part of the question worth one point, two points, or four points? If your collaborative team does not have conversations about scoring, inequities will persist across the team, and grades will not be accurate from teacher to teacher and class to class. Your team needs to articulate scoring for all exam tasks before giving the assessment instrument to students at the end of the unit.

Linking scoring points to each end-of-unit assessment task provides a means for unpacking the intent of each task to determine representations of good student work. If possible, it is important for this activity to occur before the unit begins, as the discussions will likely influence how you teach the mathematics content for the unit.

As an example, consider figure 1.25 (pages 59–63), a sample end-of-unit assessment for a possible unit on functions in grade 8.

Name: _____ Period: _____ Date: _____

Unit 2: Grade 8 Functions Assessment

Time: Sixty minutes

Tools allowed: Pencil and calculator

Content standard cluster: Define, evaluate, and compare functions.

- **8.F.1:** I can identify a relation as a function (algebraically, graphically, numerically, and verbally).

- **8.F.2:** I can compare and analyze two functions represented in different ways (algebraically, graphically, numerically, and verbally) and provide support.

1. Use the pizza menus given to answer the following questions.

Chicago Pizza Restaurant

Thin Crust	Individual 9"	Small 12"	Medium 14"
Plain Cheese	$7.25	$10.25	$12.25
Extra Ingredients	$1.00	$1.25	$1.50

Italian Pizza Restaurant

Plain Cheese Pizza . . . . . . $10.50 each

Vegetable Topping . . . . . . $1.00 each

Meat Topping . . . . $2.00 each

a. Complete the table with the number of toppings (x) and corresponding pizza cost (y) for 0–4 toppings. You may pick which size pizza to use.

Number of Toppings (x)	Cost (y)

a. Complete a table with the number of toppings (x) and corresponding pizza cost (y) for 5 toppings.

Number of Toppings (x)	Cost (y)
0 Toppings	
1 Vegetable topping	
1 Meat topping	
2 Vegetable toppings	
2 Meat toppings	

Figure 1.25: Sample grade 8 functions unit exam.

continued →

b. Graph the number of toppings (*x*) and corresponding cost (*y*) on the graph below. Be sure to label your axes.

b. Graph the number of toppings (*x*) and corresponding cost (*y*) on the graph below. Be sure to label your axes.

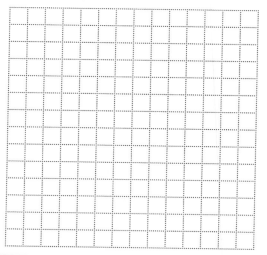

2a. Do the Chicago Pizza Restaurant data represent a function? Y or N (Circle one.)

b. Explain how you know the Chicago Pizza Restaurant does or does not represent a function based on the table.

c. Explain how you know the Chicago Pizza Restaurant does or does not represent a function based on the graph.

3a. Does the Italian Pizza Restaurant represent a function? Y or N (Circle one.)

b. Explain how you know the Italian Pizza Restaurant does or does not represent a function based on the table.

c. Explain how you know the Italian Pizza Restaurant does or does not represent a function based on the graph.

4. Build a function rule for the restaurant or restaurants that is a function.

continued →

5. Use the pizza menus to answer the following questions.

Rosenak's Rockin' Pizza

	Cost
Cheese pizza	$5.00
Additional toppings	$1.00

Giuliano's Gooey Pizza

	Cost
Cheese pizza	$3.50
Additional toppings	$1.50

a. Based on the data, complete the tables with the number of toppings (*x*) and corresponding pizza cost (*y*) for 0–4 toppings. Be sure to label each column.

Rosenak's Rockin' Pizza

x	y

Giuliano's Gooey Pizza

x	y

b. Graph the data from both pizza restaurants on the same graph. Use one color for Rosenak's Rockin' Pizza and a different color for Giuliano's Gooey Pizza. Identify which color you chose for each. Be sure to label your axes.

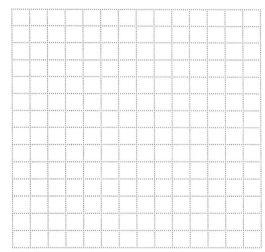

c. Determine the coordinates of the point of intersection for these two graphs. Explain what this coordinate represents in terms of both pizza restaurants.

continued →

d. If you want to spend the least amount of money, what number of toppings would make it best to order from Rosenak's Rockin' Pizza, and when would it be best to order from Giuliano's Gooey Pizza?

e. Why are there only positive *x* values?

f. Why are there only positive *y* values?

Complete each row by using the given information to complete the missing three parts. For example, in the first row, you are given a graph and will need to complete the rule, table, and verbal parts.

Rule	Table	Graph	Verbal
	<table><tr><th>x</th><th>y</th></tr><tr><td></td><td></td></tr><tr><td></td><td></td></tr><tr><td></td><td></td></tr><tr><td></td><td></td></tr><tr><td></td><td></td></tr></table>		
	<table><tr><th>x</th><th>y</th></tr><tr><td>−2</td><td>4</td></tr><tr><td>−1</td><td>1</td></tr><tr><td>0</td><td>0</td></tr><tr><td>1</td><td>1</td></tr><tr><td>2</td><td>4</td></tr></table>		

continued →

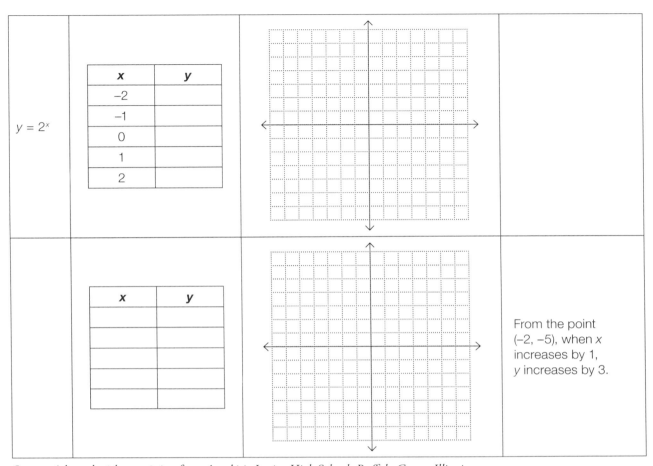

Source: *Adapted with permission from Aptakisic Junior High School, Buffalo Grove, Illinois.*

Figure 1.25: Sample grade 8 functions unit exam (continued).

Visit **go.solution-tree.com/mathematicsatwork** to download a reproducible version of this figure.

Creating the Scoring Rubric

Work with your collaborative team to assign a scoring guide of points for the end-of-unit assessment in figure 1.25. Use the assessment instrument alignment and scoring rubric tool in figure 1.26 (page 64) to answer questions for each test task.

Once you have completed the questions in figure 1.26, visit **go.solution-tree.com/mathematicsatwork** and download the actual scoring guide provided by the middle school teacher team using this assessment. The teacher team decided on eighty-five points total for the test. Does this surprise you? Did you score the test for more total points or less?

There isn't one right way to determine the points for scoring exams. What is important is that your team members use the same scoring scale and base the scoring rubrics for each task on a decided standard (such as the complexity of reasoning required by the assessment task or a proficiency scale based on lower- or higher-level cognitive demand). Each team member must also honor the agreed-on scoring scale in order to ensure grades are accurate and equitable across team members and sections of the course. This is discussed in more detail with HLTA 9 in chapter 3 (page 143).

Directions: Within your collaborative team, answer each of the following questions in relation to the assessment in figure 1.25 (pages 59–63).

1. Which essential learning standard does each task address, and how do you know that the task is aligned to the essential learning standard?

2. What work do you expect students to demonstrate in order to successfully respond to and receive full credit for each task on the assessment?

3. How will you assign partial credit for each task?

4. Which Mathematical Practices or processes does the task develop? Describe why or why not.

5. What scoring value or point value would you assign to each task?

6. Based on your scoring assignment for each task, how many total points should be assigned to this end-of-unit assessment?

7. Are there any questions on the test you would want to ask differently? If so, how would that affect the point value assigned to the test question or task?

8. How many points correct would a student need for each essential learning standard in order to be considered proficient for that standard (the proficiency target)?

Figure 1.26: Assessment instrument alignment and scoring rubric tool.

Visit **go.solution-tree.com/mathematicsatwork** to download a reproducible version of this figure.

Setting Proficiency Targets

Notice the last question in figure 1.26: How many points correct would a student need for each essential learning standard in order to be considered proficient for that standard (the proficiency target)?

Your grade-level team should decide what level of student performance will be required to be considered proficient in each of the essential learning standards for the end-of-unit assessment. Your team should know the learning score target you will expect each student to obtain for each learning standard represented on the end-of-unit test to be considered proficient for that essential learning standard. You can use the scoring rubric at **go.solution-tree.com/mathematicsatwork** or your own rubric you developed for the test in figure 1.25 to practice setting your own proficiency targets. Your team's response to students who do or do not achieve the learning proficiency target is discussed further in chapter 3 as part of HLTA 9.

Also note that there is a type of standards-based grading practice gaining popularity in grades K–8 across the United States (Reeves, 2008). It involves measuring students' proficiency on well-defined course learning standards (Marzano, 2009). Although many districts adopt standards-based grading *in addition* to traditional grades, standards-based grading can replace traditional point-based grades. If this is the case at your school, and you want more information on standards-based grading, you can go to www.marzanoresearch.com to review Robert Marzano's (2009) *Formative Assessment and Standards-Based Grading* or *A School Leader's Guide to Standards-Based Grading* (Heflebower, Hoegh, & Warrick, 2014) and learn more about the use of proficiency scales to score student work and measure student progress.

Your Team's Progress

It is helpful to diagnose your team's current reality and action prior to launching the unit. Ask each team member to individually assess your team on the fourth high-leverage team action using the status check tool in table 1.4 (page 66). Discuss your perception of your team's progress on developing scoring rubrics and proficiency expectations for the common assessment instruments. It matters less which stage your team is at and more that you and your team members are committed to working together and understanding the various student pathways for demonstrating solutions to the mathematical tasks on your common assessments as your team seeks stage IV—sustaining.

These first four high-leverage team actions give you and your team:

- A direct focus on your unit-by-unit understanding of and decisions regarding the essential learning standards
- Insight into the mathematical tasks and activities that support your work during the unit
- Understanding of common assessments you can use to determine whether or not students have attained the content and process knowledge of the essential learning standards
- Guidelines for how to score student work and set proficiency expectations for each essential learning standard of the unit

There is one major high-leverage, equity-based team action left to complete before you launch into the unit and your instruction: planning for and using common homework assignments.

Table 1.4: Before-the-Unit-Begins Status Check Tool for HTLA 4—Developing Scoring Rubrics and Proficiency Expectations for the Common Assessment Instruments

Directions: Discuss your perception of your team's progress on the fourth high-leverage team action—developing scoring rubrics and proficiency expectations for the common assessment instruments. Defend your reasoning.

Stage I: Pre-Initiating	Stage II: Initiating	Stage III: Developing	Stage IV: Sustaining
We do not use common scoring rubrics on our assessments.	We discuss our scoring and grading practices collaboratively.	We create scoring rubrics for our common unit assessments collaboratively.	We create dependable scoring rubrics for all tasks on the common unit assessments as a collaborative team.
Each teacher establishes his or her own scoring system for their independent assessments.	We have not yet reached agreement on how to score the tasks on our common assessments.	We discuss and reach agreement on a student's complete response to receive full credit on each task for our common assessments.	We design assessment rubrics to align with students' reasoning about the mathematics for each essential learning standard of the unit.
We do not know the scoring and grading practices other members of our team use.	We use scoring rubrics independently, and do not discuss our use of scoring rubrics with other members of the team.	We use the common end-of-unit assessment scoring rubrics for measuring student proficiency on each learning standard but don't discuss them as a team.	We use the common end-of-unit assessment scoring rubrics for measuring student proficiency on each learning standard and discuss them as a team.
We do not set student proficiency targets for each essential learning standard of the unit.	We set student proficiency targets independently, but do not know the proficiency targets other members of our team use for each essential learning standard of the unit.	We collaboratively set student proficiency target performances on the end-of-unit assessment for some, but not all, of the essential learning standards of the unit.	We collaboratively set student proficiency target performances on the end-of-unit assessment for each essential learning standard of the unit.

Visit **go.solution-tree.com/mathematicsatwork** to download a reproducible version of this table.

HLTA 5: Planning and Using Common Homework Assignments

Assign work that is worthy of their best effort (problem solving and reasoning).

—Linda Darling-Hammond

By using homework for practice in self-assessment and complex thinking skills, we can put students in charge of the learning process.

—Cathy Vatterott

Planning common homework assignments is another way your team reaches agreement on the second critical question of a PLC, How will we know if they know it?

High-Leverage Team Action	1. What do we want all students to know and be able to do?	2. How will we know if they know it?	3. How will we respond if they don't know it?	4. How will we respond if they do know it?
Before-the-Unit Action				
HLTA 5. Planning and using common homework assignments	▣	▢	▣	▣

▨ = Fully addressed with high-leverage team action

▭ = Partially addressed with high-leverage team action

The What

The mathematical tasks and problems you assign as homework should help your students accurately answer the question, How will I know if I am understanding the daily learning objectives from the lesson? Although not always interpreted this way, when implemented effectively, homework can become one of the most effective daily formative assessment tools available to your team as you work to continually "elicit and use evidence of student thinking"—one of the eight research-informed instructional practices outlined by NCTM (2014) in *Principles to Actions*. Thus, your grade-level or course-based team needs to reach agreement on the purpose, coherence, rigor, and length of homework assignments for every unit throughout the year. In addition, your team needs to agree on how the homework will be used and communicated to students, parents, and support staff.

Why is this an important before-the-unit-begins high-leverage team action? Once again, your team's work to develop common homework assignments for the unit before it begins becomes a potential inequity eraser for you and your students. Also, mathematics homework in middle school is often an area that lacks clarity and purpose for students, parents, intervention support personnel, and most importantly, you. Your team asks, "Why do we give students homework? What is the purpose of homework? Why won't students do their homework? How is homework assigned for a grade?" The very idea of mathematics homework in middle school, and what to do with it, is often a conundrum.

Is homework really an essential element to the process of student learning? The short answer is yes, but the best protocols to follow for homework are not quite as clear. What is clear is that:

1. The assignment of independent practice or homework cannot be a superficial exercise for you or your team.

2. Anyone who is an expert at anything devotes significant time to practice (Gladwell, 2008).

3. If we deny students an opportunity for independent practice, we deny them the very thing they need to develop real competence (Anderson, Reder, & Simon, 1995).

The homework you assign, as well as the way you think about homework as a class activity—the way you use it as a formative task to guide instruction—needs to be a carefully thought out and planned for team discussion, agreement, and activity *before* the unit begins.

Although research on homework does not indicate a specific set of common implementation protocols for all grades and all subjects (Cooper, 2008a, 2008b), the issue of homework becomes more complicated as your attention turns to implementing mathematics standards that focus on understanding (and thus using higher-level-cognitive-demand practice problems).

Although there are several schools of thought about the role homework should play and the extent of its use, research does indicate that homework can be helpful in improving student achievement if implemented correctly (Cooper, 2008b). A key finding from the research is that homework is most effective when teachers provide feedback to students' homework on a daily basis and give students written descriptive feedback that goes beyond simply marking student work as correct or incorrect (Davies, 2007; Marzano, 2007; Shuhua, 2004).

Practice is important but not without first developing student understanding. Practice without understanding may be detrimental to students' development of fluency, and in many cases, avoiding this danger means that instruction should place greater emphasis on guided practice—practice that is supported by monitoring and feedback—prior to independent practice (Larson, 2011). Marzano (2007) finds that to have a positive effect, homework should also have a clear purpose that you communicate to students: to deepen students' conceptual understanding, enhance their procedural fluency, or allow them the opportunity for independent formative practice around higher-level-cognitive-demand tasks. You should intentionally consider and carefully choose each homework problem or mathematical task based on the essential learning standards of the lesson and unit.

Research also supports the idea of *spaced* (sometimes called *distributed* or *spiral*) versus *massed* homework practice during the unit of study (Hattie, 2012; Pashler, Rohrer, Cepeda, & Carpenter, 2007) as having a significant impact on student learning. That is, provide homework assignment (practice) tasks that are spaced throughout the unit, allowing your students to cycle back and perform distributed practice on prior learning standards, including those learned earlier in the unit, previous units, or possibly in a previous course.

As each teacher on your team begins to honor high-leverage team actions 1 to 4 (teaching to the same set of essential learning standards and designing high-quality common assessments) for your course, then, it is a natural outcome that the nature of practice for student learning *outside of class* (homework) would be designed from the same core set of problems for each student, no matter the assignment of teacher for the course.

The How

Your collaborative team discussion regarding the role of homework and the selection of homework problems can be a powerful professional growth experience and should be an embedded part of your team's work throughout the year.

Understanding the Purpose of Homework

Understanding the purpose of mathematics homework on a daily basis during each unit is your first step to significantly improving current homework practice. Use the questions in the discussion tool in figure 1.27 (page 70) to help you and your team develop a better understanding of the purpose, content, and expected protocols for the unit's homework assignments. You can also use the prompts for team discussion with vertical course-based teams as you examine mathematics homework protocols and progressions across all courses in your department.

Your answers to the questions in figure 1.27 will likely vary a bit for each of your collaborative teams. It is the expectation, however, that your collaborative team will reach full agreement on your responses to the questions in figure 1.27 as you work together (before the unit begins) to select appropriate independent practice tasks (homework problems) for students to do outside of the classroom.

Your responses as individuals and as a team to these questions will reveal some of your current beliefs about assigning mathematics homework. In response to question one in figure 1.27—Why do we assign homework for each unit's lessons? What is the purpose of homework?—it is important to note that the primary purpose of homework is not summative; you should rarely assign homework to students in order to assign a grade. In fact, homework should generally not count for more than 5 to 10 percent of the total student grade. Because homework is a *formative learning* activity—an opportunity for students to obtain independent feedback and improve learning without you guiding them—it should not constitute so much of a student's grade that it is not reflective of actual student performance and achievement.

> Thus, the primary purpose of mathematics homework for middle school students is to create a formative feedback process as part of *independent practice*.

More importantly, *successful* independent practice. That is, students must understand and use homework as an opportunity for a self-guided formative assessment learning process—while you are not in the room (Hattie, 2012). Independent practice can be with other students, with other adults, or with help from YouTube or other social media resources. However, your students, while outside of class and away from you as their authority for guided practice, must practice mathematics problems and connect those problems to the essential learning standards.

Students should not view homework as something to do to receive a grade or because you, as their teacher, will go over the problems in class the next day (which makes homework no longer an independent practice exercise) or because they are being punished; rather, they should view homework as important. They should view it as a type of formative assessment for successful practice critical to the learning process and to help them retain content knowledge in their long-term memory. In class, students need your modeling and a lot of peer-to-peer guided practice (see chapter 2, HLTA 7, page 99, for more details). Then, outside of class, and in a timely fashion, they need to participate in accurate independent practice with feedback (self-feedback and action or with feedback with peers)—well before they return to your class the next day.

Directions: Use the following prompts to guide discussion of the unit's homework assignments.

Purpose of homework:

1. Why do we assign homework for each unit's lessons? What is the purpose of homework?

Nature of homework:

2. What is the proper number of mathematical tasks for daily homework assigned during the unit? In other words, how much time should students spend on homework?

3. What is the proper rigor (cognitive-demand expectations) of the mathematical tasks for homework assigned during the unit?

4. What is the proper distribution of tasks for homework to ensure spaced practice (cyclical review) for our students?

5. How do our daily homework assignments align to the learning standard expectations for the unit?

6. How will we reach consensus on unit homework assignments in order to ensure coherence to the student learning and practice expectations?

Use of homework:

7. How should we grade or score homework assignments?

8. What will we do if students do not complete their homework assignments?

9. How will we go over the homework in class?

10. How will we communicate the common unit homework assignments to students, parents, and support staff?

Figure 1.27: Collaborative homework assignment protocol discussion tool.

Visit **go.solution-tree.com/mathematicsatwork** to download a reproducible version of this figure.

Perhaps the homework paradigm shift for you and your colleagues is to stop calling homework *home-work*. Certainly, students can do *independent practice* at the coffee shop or after school in a classroom, on the bus, in a hallway, in the car on the way to practice or a game, or with friends at the library. Independent mathematics practice does not have to be done at home.

Ultimately, the work of your collaborative team is to decide the role homework plays as part of your classroom activities and learning process. As you work to review and develop your team's homework protocols, consider the following guidelines.

- **Homework purpose:** The primary purpose of homework should be to allow the student the opportunity for *independent practice* on learning standards mastered in class during guided practice and small-group discourse. Homework can also provide a chance for the student to practice mathematical tasks that relate to previous learning standards or tasks that reflect prerequisite learning standards for the next unit. Homework that provides review of previous work and helps to prepare students for future work leads to improved student achievement (Cooper, 2008a).

- **Homework length:** How much time should daily homework take students to complete? How many problems should it entail? Homework should not be lengthy (Cooper, 2008b), so teachers should take care about what they assign—no more than eight to ten carefully chosen problems per day. Take into account the cognitive demand of the tasks or problems you assign. Homework tasks as a general rule should not take more than thirty to thirty-five minutes (per course) of time outside of class.

- **Homework task selection:** The homework your school curriculum or textbook includes is not necessarily appropriate for your students without some adjustments with which your team agrees. Make sure that all tasks are necessary as part of independent practice, have *spaced* practice and not *massed* practice, and align to the stated learning standards of the unit.

- **Homework answers:** There are many advantages to providing students with homework answers before the unit of instruction begins. When you provide students with answers to the homework problems, they can check their solutions against the answers, and if their end results do not match the provided answers, they can rework the problem to find their errors. In other words, students receive immediate and formative self-assessed feedback of their work—like when playing an electronic game. Moreover, a compelling reason to provide students with the answers to the homework in advance of the assignment is to save time during the class period the next day. *No time* should be spent going over the answers or the actual homework problems. Remember, homework is *independent* practice, not *in-class* practice. Since the students know exactly what they know and what they do not understand, any in-class discussion time on homework can be limited to a brief few minutes and becomes more meaningful for the students.

- **Homework focus in class:** Once your collaborative team determines homework, focus on how to address homework in class, the type of feedback that teachers will give students, and what will occur if students do not complete the homework. If you spend most of the class time going over homework, you lose the impact of successful independent practice on student learning. Your students may be choosing to wait to do homework problems because they know they can write down the work when you go over the problems the next day. Since the purpose of homework is independent practice, limit the amount of time in class to grade, score, or go over the practice problems. If you spend most of the class time going over homework, your team must revisit the amount and content of what you assign. It could be that your team assigned too much homework or that students did not achieve an appropriate level of mastery prior to practice of the learning standard.

These daily and unit processes and procedures should be agreed on and consistent from teacher to teacher within your grade-level or course-based collaborative team.

Using Effective Homework Protocols

From a rigor and coherence point of view, the homework you assign for a unit of study must be the same for all students in your course. Give all team-developed homework assignments to your students and parents in advance of teaching the unit with the understanding that your team can and will modify the assignments during the unit as necessary to address specific student learning needs. Use the diagnostic tool in figure 1.28 to check how your team is doing with respect to using high-quality mathematics homework protocols and procedures.

Figure 1.29 (page 74) is a sample unit homework assignment sheet from the Algebra 1 Team at Aptakisic Junior High in Buffalo Grove, Illinois. Students receive the homework assignment sheet at the beginning of a unit, and it's also available online as a Google Doc. The team blends a use of the textbook with its own extensive worksheets for each homework assignment and posts them online as well, as the teachers do not rely on the textbook only for all assignments. You can use elements of the diagnostic tool from figure 1.28 to score the quality of the sample homework shown in figure 1.29. Note that the homework assignment sheet lists the unit's essential learning standards and that the problems assigned illustrate spaced practice that is focused. In general, these teachers spend no more than five minutes of class time going over the homework in class the next day since they view homework as independent practice for their students.

High-Quality Homework Indicators	Description of Level 1	Requirements of the Indicator Are Not Present	Limited Requirements of This Indicator Are Present	Substantially Meets the Requirements of the Indicator	Fully Achieves the Requirements of the Indicator	Description of Level 4
The primary purpose of homework is independent practice.	Homework is primarily assigned to give a student a grade. Homework counts more than 10 percent of a student's total grade.	1	2	3	4	Homework is understood as primarily for independent practice and a formative assessment learning loop for students. Homework counts no more than 10 percent of a student's grade.
Homework assignments are the same for every teacher on the course team.	Each teacher on the team creates his or her own homework assignments and does not share with others.	1	2	3	4	Common homework assignments are developed collaboratively by the team and are the same for all students in the grade level or course.
All homework assignments for the unit are given to the students before the unit begins.	Students find out homework assignments each day or each week as the unit progresses.	1	2	3	4	Students are provided all unit homework assignments—electronically or with a handout—as the unit begins.
Homework assignments for the unit are appropriately balanced for cognitive demand.	Homework practice problems are not balanced for rigor. Emphasis is on lower-cognitive-demand tasks.	1	2	3	4	Homework practice is appropriately balanced with higher- and lower-cognitive-demand tasks.
All practice problem answers are given to the students in advance of the homework assignments.	Students must wait until the next day to receive answers or solutions to homework practice problems.	1	2	3	4	Students are able to check their solutions during independent practice and are expected to rework the problems if not correct the first time.
Homework assignments for each unit exhibit spaced and massed practice.	The homework assignments represent superficial thought as to the problems chosen and consist of massed practice.	1	2	3	4	The homework assignments represent carefully chosen problems or tasks. Spaced practice from several lessons of the unit or previous units is included in addition to massed practice.
Daily homework is aligned to the essential learning standards of the unit.	Students are not able to make connections between the daily homework practice problems and the learning standards of the unit.	1	2	3	4	Students connect the homework practice as essential to helping them demonstrate knowledge of the essential learning standards of the unit.
Limited time is spent going over homework in class.	Students and teacher spend fifteen to twenty-five minutes (or more) in class going over the homework answers and solutions. The teacher does most of the work as the students watch.	1	2	3	4	At most, five to seven minutes of class time are used discussing the homework. It is primarily a peer-to-peer class activity facilitated by the teacher.

Figure 1.28: Homework quality diagnostic tool.

Visit **go.solution-tree.com/mathematicsatwork** to download a reproducible version of this figure.

Grade 8 Algebra 1—Assignment Guide Unit 1		
1. Write out the original problem. 2. Show all steps needed to solve the problem. 3. Write and box your answer. 4. Check and self-correct your answers. Use the solution key posted online as needed to check your answers.		
Lesson	**Assignment**	**Score**
Lesson One: Solving Problems by Creating Equations in One Variable (A-CED.1)	Homework worksheet one	
Lesson Two: Solving Problems by Creating Equations in One Variable (A-CED.1)	Homework worksheet two	
Lesson Three: Solving Problems by Creating Equations in One Variable (A-CED.1)	Homework worksheet three (Include a review from worksheets one and two.)	
Lesson Four: Solving Problems by Creating Equations in One Variable (A-CED.1) and Introduction to the Graphing Calculator	Homework worksheet four	
Quiz One on A-CED.1		
Lesson Five: Introduction to Functions (F-IF.1–2)	Pages 259–261 (1, 3, 13, 23, 38, 40–42)	
Lesson Six: Evaluating Functions (F-IF.1–2)	Homework worksheet five	
Lesson Seven: Operations With Functions (F-IF.1–2)	Page 260 (27, 43), page 267 (5, 29), page 271 (4, 19, 30)	
Quiz Two on F-IF.1-2		
Lesson Eight: Solving by Creating Equations in Two Variables (A-CED.2)	Homework worksheet six	
Lesson Nine: Solving by Creating Equations in Two Variables (A-CED.2)	Homework worksheet seven	
Lesson Ten: Solving by Creating Equations in Two Variables (A-CED.2)	Homework worksheet eight	
Review	Review project for the unit.	
Unit 1 Test on A-CED.1–2 and F-IF.1–2		

Source: Adapted with permission from Aptakisic-Tripp CCSD 102, Buffalo Grove, Illinois.
Source for standards: NGA & CCSSO, 2010, pp. 65, 69.

Figure 1.29: Sample unit homework assignment sheet: Algebra 1 Team, Aptakisic Junior High.

Take some time to examine your current homework assignments and practices. Use the homework quality diagnostic tool from figure 1.28 (page 73) to evaluate your current homework practices and make decisions about how your team can improve in this critical formative assessment process for students.

Your Team's Progress

It is helpful to diagnose your team's current reality and action prior to launching the unit or chapter. Ask each team member to individually assess your team on the fifth high-leverage team action using the status check tool in table 1.5. Discuss perceptions of your team's progress on planning and using common homework assignments. It matters less what stage your team is at and more that you and your team members are committed to collaboratively defining the purpose of homework, using the same common homework assignments and protocols, and communicating those assignments to students, parents, and colleagues as your team seeks stage IV—sustaining.

Table 1.5: Before-the-Unit-Begins Status Check Tool for HLTA 5—Planning and Using Common Homework Assignments

Directions: Discuss your perception of your team's progress on the fifth high-leverage team action—planning and using common homework assignments. Defend your reasoning.			
Stage I: Pre-Initiating	**Stage II: Initiating**	**Stage III: Developing**	**Stage IV: Sustaining**
We do not have a clear purpose for why we assign homework.	We have *established* a clear purpose for homework, but it is not independent and formative student practice.	We have *developed* the shared purpose of using homework as independent formative student practice.	We have *implemented* the shared purpose of homework as independent formative student practice.
We do not plan or use common homework assignments and do not know the homework assignments given by other members of our team.	We discuss homework assignments and have not yet reached collaborative agreement on the nature of those assignments for each unit.	We collaboratively *plan* and develop common homework assignments for each unit.	We collaboratively *use* common homework assignments for each unit.
We do not know the nature of the homework protocols used for the assignments given by other members of our team.	We discuss the nature of the homework protocols used for the assignments given by other members of our team, but do not agree on those protocols.	We have team agreement on developed homework protocols including limited number of tasks, spaced practice, balance of cognitive demand, and alignment to the essential learning standards.	We have complete team agreement on homework protocols including limited number of tasks, spaced practice, balance of cognitive demand, and alignment to the essential learning standards, and we use those protocols with our students.
We do not know how other members of our team go over homework in class.	We discuss how we go over homework in class but do not agree on what we should do.	We discuss how we go over homework in class and agree on what we should do with homework during class.	We discuss how we go over homework in class, agree on what we should do, and implement that agreement.
We do not know how other members of our team count homework as a percent of the student's total grade.	We know how others count homework for a grade, but we each do it our own way.	We grade homework the same each day, but we count it differently from other team members as a percent of the total student grade.	We have complete team agreement on how homework should be used and accounted for as part of the student's total grade.

Visit **go.solution-tree.com/mathematicsatwork** to download a reproducible version of this table.

As your team seeks the sustaining stage, you will increase the rigor, coherence, and fidelity of the independent practice (homework) all students are expected to do during the unit for your course.

Setting Your Before-the-Unit Priorities for Team Action

When your school functions within a PLC culture, your grade-level team makes a commitment to reach agreement on the five before-the-unit-begins high-leverage team actions outlined in this chapter.

> HLTA 1. Making sense of the agreed-on essential learning standards (content and practices) and pacing
>
> HLTA 2. Identifying higher-level-cognitive-demand mathematical tasks
>
> HLTA 3. Developing common assessment instruments
>
> HLTA 4. Developing scoring rubrics and proficiency expectations for the common assessment instruments
>
> HLTA 5. Planning and using common homework assignments

As a team, reflect together on the stage you identified with for each of these five team actions. Based on the results, what should be your team's priority? Use figure 1.30 to focus your time and energy on actions that are most urgent in your team's preparation for the next unit. You and your team cannot focus on everything. Focus on fewer things, and make those things matter at a deep level of implementation.

The five high-leverage team actions in this chapter combine to form step one of the teaching-assessing-learning cycle (see figure 1.1, page 8) and will help you prepare for the rigors and challenges of teaching and learning during the unit. They are also linked to teacher actions that will significantly impact student learning in your class.

In chapter 2, we turn our attention to steps two and three of the teaching-assessing-learning cycle, which focus on implementing formative assessment classroom strategies and students taking action on in-class formative assessment feedback. We also focus on supporting student engagement in the Mathematical Practices to promote deeper understanding of mathematical content through the use of higher-level-cognitive-demand tasks. The CCSS Mathematical Practices lesson-planning tool provides one avenue for organizing your collaborative team's work for collective lesson inquiry.

Directions: Identify the stage you rated your team for each of the five high-leverage team actions, and provide a brief rationale. When you are ready, discuss your ratings as a team.

1. Making sense of the agreed-on essential learning standards (content and practices) and pacing

   Stage I: Pre-Initiating   Stage II: Initiating   Stage III: Developing   Stage IV: Sustaining

   Reason: _____

   _____

   _____

2. Identifying higher-level-cognitive-demand mathematical tasks

   Stage I: Pre-Initiating   Stage II: Initiating   Stage III: Developing   Stage IV: Sustaining

   Reason: _____

   _____

   _____

3. Developing common assessment instruments

   Stage I: Pre-Initiating   Stage II: Initiating   Stage III: Developing   Stage IV: Sustaining

   Reason: _____

   _____

   _____

4. Developing scoring rubrics and proficiency expectations for the common assessment instruments

   Stage I: Pre-Initiating   Stage II: Initiating   Stage III: Developing   Stage IV: Sustaining

   Reason: _____

   _____

   _____

5. Planning and using common homework assignments

   Stage I: Pre-Initiating   Stage II: Initiating   Stage III: Developing   Stage IV: Sustaining

   Reason: _____

   _____

   _____

With your collaborative team, respond to the red light, yellow light, and green light prompts for the high-leverage team actions that you and your team believe are most urgent.

Red light: Indicate one activity you will stop doing that limits effective implementation of each high-leverage team action.

Yellow light: Indicate one activity you will continue to do to be effective with each high-leverage team action.

Green light: Indicate one activity you will begin to do immediately to become more effective with each high-leverage team action.

Figure 1.30: Setting your collaborative team's before-the-unit priorities.

Visit **go.solution-tree.com/mathematicsatwork** to download a reproducible version of this figure.

CHAPTER 2

During the Unit

The choice of classroom instruction and learning activities to maximize the outcome of surface knowledge and deeper processes is a hallmark of quality teaching.

—Mary Kennedy

Learning is experience. Everything else is just information.

—Albert Einstein

Much of the daily work of your collaborative team occurs during the unit of instruction. This makes sense, as it is during the unit that you place much of your collaborative team effort put forth before the unit into action.

Your team conversations during the unit should focus on sharing evidence of student learning, discussing the effectiveness of lessons or activities, and examining the ways in which students may be challenged or need scaffolding to engage mathematically. While discussion about some of the tasks and the end-of-unit assessment planning take place prior to the start of the unit, teachers often plan and revise day-to-day unit lessons *during the unit* as they gain information regarding students' needs and successes. What your students do and say while developing understanding of the essential learning standards for the unit provides the data for your teacher team conversations.

This process of data gathering, sharing, providing feedback, and taking action regarding student learning forms the basis of an in-class *formative assessment process* throughout the unit. By engaging in these actions, your collaborative team can make needed adjustments in task development and instruction that will better support student learning during the unit. An effective formative assessment process also empowers students to make needed adjustments in their ways of thinking about and doing mathematics to lead to further learning. These actions support NCTM's (2014) assessment principle and the research-informed instructional practice that calls for teachers to "elicit and use evidence of student thinking" to advance student learning. They also support NCTM's (2014) teaching and learning principle that engages students in meaningful learning through individual and collaborative experiences that promote their ability to make sense of mathematical ideas and reason mathematically by implementing tasks that promote reasoning and problem solving and support productive struggle in learning mathematics. See appendix E (page 189) for more details about how each HLTA supports NCTM's principles outlined in *Principles to Actions*.

This chapter is designed to help you and your collaborative team members prepare and organize your team's work and discussions around three high-leverage team actions during the unit of instruction. These three high-leverage actions support steps two and three of the PLC teaching-assessing-learning cycle in figure 2.1 (page 80).

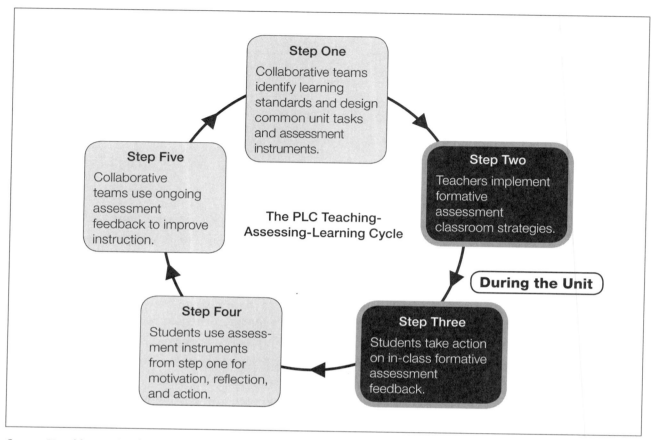

Source: Kanold, Kanold, & Larson, 2012.

Figure 2.1: Steps two and three of the PLC teaching-assessing-learning cycle.

The three high-leverage team actions that occur during the unit of instruction are:

> HLTA 6. Using higher-level-cognitive-demand mathematical tasks effectively
>
> HLTA 7. Using in-class formative assessment processes effectively
>
> HLTA 8. Using a lesson-design process for lesson planning and collective team inquiry

Steps two and three of the teaching-assessing-learning cycle provide a focus to your collaborative team's use of effective in-class strategies that support students' mathematics learning during the unit's instruction. How well your collaborative team implements meaningful formative feedback and assessment processes (step two) is only as effective as how well you and your team are able to elicit student actions and responses to the formative feedback you and their peers provide (step three). You can only effectively implement formative assessment if students are actively involved in the feedback process.

As you begin your during-the-unit team discussions, set the stage by developing a common understanding of key terms you will use. You and your team should complete table 2.1 (page 81) to organize your thinking prior to continuing through this chapter. Each team member will present common as well as different ways of thinking about teaching, assessing, and learning. A common understanding of the terms provides a foundation that facilitates thorough consideration of mathematics instruction to maximize student achievement. Once again, this work is about the constant and ongoing discussion between you and your team.

Table 2.1: Common Understandings of Key Terms

Directions: Record your understandings of the following key terms. Provide an example to illustrate your understandings, and compare and contrast your understandings with other team members.		
Term	**Your Understanding**	**Example to Illustrate Your Understanding**
Teaching mathematics		
Assessing mathematics		
Learning mathematics		
Checking for understanding		
Using formative assessment processes		

Visit **go.solution-tree.com/mathematicsatwork** to download a reproducible version of this table.

After each team member has responded to table 2.1, discuss how your team can use these key terms during a unit of instruction to bolster students' mathematics learning through collaborative team efforts. For example, some might include *student learning* in their definition of *teaching*, and others might not have considered that teaching is not effective unless a recipient of that teaching actually learns. This should lead to a discussion of how team members interpret the term *teaching*. You should revisit and consider understandings such as this example as you work your way through this chapter. It is important to align your understandings of these terms to pursue an equitable formative assessment process.

As you focus attention on Mathematical Practices and processes as an integral part of your instruction, the challenge is to envision these practices as student outcomes in the classroom. As you collaborate with your colleagues around instruction, your dialogue will focus specifically on the tasks you use, the questions you ask and students answer, the nature of your whole- and small-group discourse, and the way you manage the daily activities in which students participate. This should lead your team to consider the question, What are students doing as they engage in the Mathematical Practices and processes?

The three high-leverage team actions in this chapter will allow you to reach higher levels of student achievement in your class than ever before. Your students will go deeper in their use of higher-level-cognitive-demand tasks, demonstrating the Mathematical Practices, using formative assessment processes, and understanding the essential learning standards in the middle school standards, including groupings of the standards around content standard clusters.

HLTA 6: Using Higher-Level-Cognitive-Demand Mathematical Tasks Effectively

This sixth high-leverage team action highlights the team's work to present, adjust, and use daily common higher- and lower-level-cognitive-demand mathematical tasks. These tasks were designed in step one of the teaching-assessing-learning cycle as part of your HLTA 2 (page 22) work.

Recall there are four critical questions every collaborative team in a PLC culture asks and answers on an ongoing unit-by-unit basis.

1. What do we want all students to know and be able to do? (The essential learning standards)

2. How will we know if they know it? (The assessment instruments and tasks teams use)

3. How will we respond if they don't know it? (Formative assessment processes for intervention)

4. How will we respond if they do know it? (Formative assessment processes for extension and enrichment)

The sixth HLTA—using higher-level-cognitive-demand mathematical tasks effectively—ensures your team reaches clarity on the second PLC critical question, How will we know if they know it?

High-Leverage Team Action	1. What do we want all students to know and be able to do?	2. How will we know if they know it?	3. How will we respond if they don't know it?	4. How will we respond if they do know it?
During-the-Unit Action				
HLTA 6. Using higher-level-cognitive-demand mathematical tasks effectively	▨▢	▨		

▨ = Fully addressed with high-leverage team action

▨▢ = Partially addressed with high-leverage team action

The What

You and your team intentionally plan for, design, and implement higher-level-cognitive-demand mathematical tasks that will provide ongoing student engagement and opportunities for descriptive feedback to your students. This high-leverage team action supports the essential learning standards for the unit as well as Mathematical Practices and the mathematical content.

Effective use of higher-level-cognitive-demand tasks means you do not lower the task's cognitive demand during instruction. When students work on cognitively demanding tasks, they often struggle at first. Some mathematics teachers perceive student struggle as an indicator that they have failed instructionally. Thus, the teacher is tempted to rescue students by breaking down the task and guiding students step by step to a solution (NCTM, 2014). This, in turn, deprives students of the opportunity to make sense of the mathematics (Stein et al., 2007). The critical need to effectively implement mathematical tasks without lowering the cognitive demand of the task, along with strategies to avoid cognitive decline, are outlined in the research-informed instructional practice, "support productive struggle in learning mathematics" in *Principles to Actions* (NCTM, 2014, p. 48).

HLTA 6 consists of the intersection of three action components:

1. Understanding student proficiency in Mathematical Practices and processes
2. Maintaining a high level of cognitive demand for the mathematical tasks you use in class
3. Maintaining focus on the essential learning standards during the unit of instruction

By emphasizing these three components, you and your collaborative team can dissect this high-leverage team action and develop plans for making it a reality. The process begins by revisiting and exploring what it means for students to be proficient with the Standards for Mathematical Practice: practices that describe *how* students should engage with the mathematics. To review, the eight Standards for Mathematical Practice (presented in more detail in appendix A, page 177) are (NGA & CCSSO, 2010, pp. 6–8):

1. Make sense of problems and persevere in solving them.
2. Reason abstractly and quantitatively.
3. Construct viable arguments and critique the reasoning of others.
4. Model with mathematics.
5. Use appropriate tools strategically.
6. Attend to precision.
7. Look for and make use of structure.
8. Look for and express regularity in repeated reasoning.

The How

To begin your work on this high-leverage team action, spend time deeply exploring each of the eight Mathematical Practices with your team. These standards represent important processes for student learning, whether your state is participating in the Common Core standards or not. In *Common Core Mathematics in a PLC at Work, Grades 6–8* (Kanold et al., 2013), we used three key questions to help you and your team better understand the Standards for Mathematical Practice.

1. What is the intent of the Mathematical Practice, and why is it important?
2. What teacher actions facilitate student engagement in this Mathematical Practice?
3. What evidence exists that students are demonstrating this Mathematical Practice?

Each team member—and your students—should be able to respond accurately and with depth to these three questions. Your team can use the Frayer model (figure 2.2) as an effective technique to investigate the intent and reasoning behind each Mathematical Practice. This model provides a useful framework to unpack the meaning of the Mathematical Practices with a focus on classroom implementation. You and your collaborative team should work through each of the eight practices one at a time to create posters (electronically or with poster paper) using the Frayer model. Post these in your department office area as a reminder of student expectations for demonstrating each Mathematical Practice. What does each practice look like and sound like in your classroom? That is, what should you expect to see and hear from your students when they are doing mathematics in your classroom?

If your district's standards are not Common Core then use either the mathematics process standards for your state or the effective teaching practices from NCTM (2014).

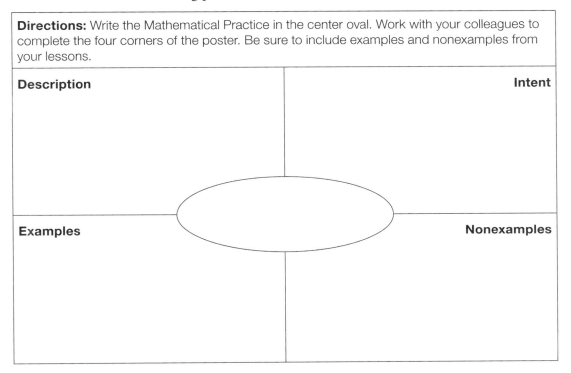

Source: Adapted from Frayer, Fredrick, & Klausmeier, 1969.

Figure 2.2: Using the Frayer model for the Mathematical Practices.

Since the Mathematical Practices represent what students are to do, one of your first responsibilities is to expect students to demonstrate their understanding of each process and standard as the year progresses. Your second responsibility is to ensure students experience mathematical tasks or activities that allow them to actually demonstrate the Standards for Mathematical Practice as part of your lesson planning for each unit.

Likely, your examples and nonexamples will reflect what you currently see, or plan to see, in your classroom around each Mathematical Practice. For example, a Frayer model poster for Mathematical Practice 5, "Use appropriate tools strategically," might have information related to the following for its description, intent, example, and nonexample:

- **Description**—Students know what tools are useful for given problems and use those tools in ways to increase efficiency and understanding. Students know the benefits of using one tool over another for a given problem context and discern appropriately when to use different tools.

- **Intent**—Students have access to a wide variety of tools. The tools students choose provide data in the formative assessment process related to how students think about the problem.

- **Example**—Based on the nature of the coefficients generated by real-life modeling of population data, middle school students might use graph paper (online or paper) and a graphing calculator tool as a way to determine a visual representation for a solution to a system of linear equations. Students could also use an app to verify the solution through the use of a built-in "intersect" function that allows identification and interpretation of the points of intersection, if any.

- **Nonexample**—The teacher does not connect the solution for a system of equations to the graph, does not provide for tasks representing mathematical models of the standard, and directs the students to not use any tools to support student understanding and reasoning.

Discuss your team's examples and nonexamples from the Frayer model activity to determine if the examples are present in your classrooms. Discuss what your team can do to make the focus on Mathematical Practices and processes more common during your lessons. You can use the ideas from your team discussion as a basis for understanding how to capture student proficiency in these Mathematical Practices. See appendix B (page 181) for additional ideas.

Understanding Student Proficiency in the Mathematical Practices

It is important to note that your team's investigation of Mathematical Practices and processes is all about understanding how students are to *learn and do* mathematics. As a reader and user of this handbook, whether or not your state adopted the Common Core, is a member of an assessment consortia, or has established individual state standards and assessments, is not essential to this high-leverage action. Research about how students learn mathematics at high levels of achievement is the basis on which Mathematical Practices and processes stand (Hattie, 2012; Kilpatrick, Swafford, & Findell, 2001; NCTM, 2007, 2014). In short, your deep understanding of how to develop student proficiency in these practices and processes has a considerable learning benefit to the student. (See appendix D, page 187, or visit **go.solution-tree.com/mathematicsatwork** to access research resources related to the ten high-leverage team actions.)

You and your colleagues will need to develop consensus on the meaning of *proficiency* relative to students' engagement with Mathematical Practices and processes in middle school. To support this work, use figure 2.3 to organize your collaborative team's initial ideas about proficiency and the Mathematical Practices. For an example of a team brainstorm, see the evidence list of each Mathematical Practice in appendix B (page 181). You can also use the resources listed in appendix D (page 187) to help you with this team activity.

Directions: Record your insights about proficiency with these Standards for Mathematical Practice and processes.
Mathematical Practice 1: When students are proficient with making sense of problems and persevering in solving them, they . . .
Mathematical Practice 2: When students are proficient with reasoning abstractly and quantitatively, they . . .
Mathematical Practice 3: When students are proficient with constructing viable arguments and critiquing the reasoning of others, they . . .
Mathematical Practice 4: When students are proficient with modeling with mathematics, they . . .
Mathematical Practice 5: When students are proficient with using appropriate tools strategically, they . . .
Mathematical Practice 6: When students are proficient with attending to precision, they . . .
Mathematical Practice 7: When students are proficient with looking for and making use of structure, they . . .
Mathematical Practice 8: When students are proficient with looking for and expressing regularity in repeated reasoning, they . . .

Figure 2.3: Identifying proficiency with the Standards for Mathematical Practice.

Visit **go.solution-tree.com/mathematicsatwork** to download a reproducible version of this figure.

Students' proficiency with the Mathematical Practices will be apparent through effective class discussion and instruction around both lower- and higher-level-cognitive-demand tasks. The lower-level-cognitive-demand tasks might involve teacher-student dialogue; however, the higher-level-cognitive-demand tasks must include student-to-student discussions about the tasks.

Using Higher-Level-Cognitive-Demand Mathematical Tasks Effectively During the Unit of Instruction

Consider the activity and scenario presented in figure 2.4 and figure 2.5 (pages 89–90). With your team, respond to the questions in figure 2.4 once you read the scenario, and as you examine the inequality task activity from figure 2.5.

Directions: Read the scenario, and examine the inequality task activity described in figure 2.5 (pages 89–90). Then, answer the questions that follow the scenario.

Ms. Schneider enters her sixth-grade accelerated class prepared to facilitate instruction to engage her students in an active learning experience. A few of her team members were able to come to her class and observe students during the lesson. As her students are settled into the class, she presents the inequality task shown in figure 2.5 (pages 89–90). Since she had planned the unit with the sixth-grade team, she is excited about sharing her students' experiences when the team meets again.

1. Ms. Schneider and her team created this task because they thought it would engage students. Why do you suppose they considered the inequality task an engaging task?

2. How would you modify the task to increase its potential to engage students?

3. What is the essential learning standard for this mathematical task?

4. What might make the mathematics of the task less or more challenging for your students?

5. What is the real-life model for this task? What other real-life model might be a useful context for the same mathematics in this task?

6. What can you learn about students as they engage in and respond to this task?

7. What do you expect students' questions to be regarding the task? What do you expect students' misunderstandings to be regarding the task?

8. Because student responses to the mathematical task will vary, how do you propose team members use the task in class?

Figure 2.4: Mathematics instruction and higher-level-cognitive-demand task scenario.

Visit **go.solution-tree.com/mathematicsatwork** to download a reproducible version of this figure.

Essential learning standard: 7.EE.4b—I can write an inequality of the form $px + q > r$ or $px + q < r$ to represent a constraint in a real-world context using $<$, $>$, $\leq$, or $\geq$.

Take Flight Aircraft, Inc.
Hutchinson Municipal Airport
(888) 555-4545

500 Airport Road
Hutchinson, KS 67501

Rental Aircraft Prices

Aircraft Rates	
Cirrus SR20 (N620DA)	$195.00 per Hour (wet)
Cessna Skyhawk SP (N349SP)	$125.00 per Hour (wet)
Headset Rental	$3.00 per Flight
All aircraft rental rates include fuel costs (wet).	

Flight Training		Other Fees & Rates	
Ground Instruction	$40.00/Hour	Private Pilot Kit	$220.00
Flight Instruction (Primary)	$40.00/Hour	Instrument Pilot Kit	$165.00
Flight Instruction (Advanced)	$50.00/Hour	Commercial Pilot Kit	$165.00
Flight Instruction	$350.00/Day		
Prices subject to change.			

You are looking to take airplane lessons. The graphic shows all the prices from Take Flight Aircraft, Inc. You have been saving your money and have a total of $3,750 to spend on your new hobby.

1. The first step for your training includes ground instruction, but you are unsure of the number of hours you will need, as every pilot is different. Your budget for the ground instruction is one-fourth of your total budget. You also need to purchase the private pilot kit.

   Set up an inequality that represents the first step to your flight instruction, and then solve the inequality.

2. Once your instructor has cleared you from your ground instruction, you must begin your primary instruction. With your primary instruction, you must rent an airplane. The airplane comes with fuel (wet), so there is no need to worry about the cost for fuel, but you are unsure how many hours your instructor will need you to train as every pilot is different. Your estimated budget is one-fourth of the total budget. You must rent the headset, but there is a new deal: a $25 unlimited plan for the year. Write and solve two inequalities for your primary instruction (one involving each airplane type).

3. Next, your instructor will move you on to the advanced class. You must rent the headset again, but it is under the same plan. The remaining budget is saved for your advanced instruction, as this is a longer part of the training. Write and solve two different inequalities for your advanced instruction (two different types of planes).

You passed and would like to plan your first trip with one of these aircrafts! You need to do some research about your plane and decide on your destination, which must have an airport. You will be taking off from Chicago Executive Airport in Wheeling, Illinois, or Schaumburg Airport in Schaumburg, Illinois.

Figure 2.5: Inequality task.
continued →

On a separate page or using Pic Collage (an app that allows you to write or import images), include the following:

Your precalculations:

Most small aircrafts have two gas tanks with 20 gallons in each (total of 40 gallons). Usually when flying, there are a lot of factors to consider when figuring out how far or how much gas you will need (wind, altitude, weight, and so on). We will assume that our airplane will use 10 gallons of fuel per hour and you will travel at 110 mph. Remember that distance = rate • time.

Include calculations for your aircraft as well as a picture.

1. Write an inequality for cost of your flight (refer to the cost table in the graphic). Let x = travel time in hours.

2. Write an inequality for the amount of gas your airplane has in its tank (assuming it is full at take-off).

Now you need to make your actual trip plan. (I hope you have figured out that you cannot go too far in this type of plane!)

Using Google Maps:

1. Place your markers on each location, and zoom out.

2. Take a screen shot, and include this on your page.

3. Use your scale to determine how many miles your flight will be. Show your calculation. Do not round to the nearest inch or centimeter and so on. Fractions and decimals should be included.

Now make calculations for your flight:

1. How long will your flight take? (Determine to the nearest minute—no fractions or decimals of an hour.)

2. How many gallons of gas will your flight use?

3. What is the cost for your flight?

Finally, use Google search to find a model of your airplane to include on your Pic Collage.

Once you have finished your page, print it out in color, and turn it in.

Now you get to try to fly your airplane!

Using the Google Earth app:

1. On the computer, click on the Google Earth app.

2. Type in your airport location to get to your starting point.

3. Go to the Tools drop-down menu, and click on Enter Flight Simulator.

4. Use the arrow buttons on your keyboard to fly your plane.

Extension questions (fun extra work when you are finished):

Pack your bags!

Make up your dream flight. Map it on Google Maps, and calculate the cost. You will probably need to stop along the way to refuel. Gas is $5.78 per gallon.

Source: Reprinted with permission from Meridian Middle School, Buffalo Grove, Illinois.
Source for standard: NGA & CCSSO, 2010, p. 49.

Figure 2.5: Inequality task (continued).

Your effective use of higher-level-cognitive-demand tasks during the unit of instruction (to develop student reasoning and problem-solving ability) will benefit from team discussions using questions such as those in figure 2.4 (page 88). During any given unit, each team member should bring a higher-level-cognitive-demand mathematical task to the team meeting. The task can be teacher developed or from a curriculum resource, but it should be one your teacher team plans to use as part of the second high-leverage team action from chapter 1—identifying higher-level-cognitive-demand mathematical tasks. The task can take the full period, such as the airport task from figure 2.5, or it can take just ten minutes of class time. Just make sure that the tasks provide an expectation for higher-level cognitive demand for students.

In general, you can use the tool in figure 2.6 to promote deep discussion of how to use or manage *any task* you bring to the team for potential use in class. As an example, consider the mathematical task presented in figure 2.7 (page 92), and answer the questions from figure 2.6 in order to learn more about how you might use the task effectively in class.

Directions: Use the following questions to allow for deep discussion of each task.

1. What is the essential learning standard of the unit this task addresses? Why did you select (or create) this task?

2. How would you modify the task to increase its potential to engage students during use of the task?

3. What do you expect students' misunderstandings to be regarding the task?

4. What might make the mathematics of the task less or more challenging to meet the needs of students?

5. What is a real-life model of this task? What other real-life models might provide a useful context for the mathematics in this task?

6. What can you learn about students as they engage in and respond to the task?

7. What are all possible ways students can find a solution pathway for the task? What do you expect students' misunderstandings to be regarding the task?

8. How do you propose to effectively manage and facilitate the use of the task?

Figure 2.6: Mathematics instruction and higher-level-cognitive-demand task development tool.

Visit **go.solution-tree.com/mathematicsatwork** to download a reproducible version of this figure.

Essential learning standard: 8.EE.8c—I can solve real-world mathematical problems involving two linear equations in two variables.

The local swim center is making a special offer. It usually charges $7 per day to swim at the pool. This month, swimmers can pay an enrollment fee of $30 and then the daily pass will only be $4 per day.

1. Suppose you do not take the special offer. Build a function that represents the amount of money you would spend based on how many days you go to the pool if the passes were purchased at full price.

2. Build a second function that represents the amount of money you would spend if you decided to take the special offer.

3. Graph your two functions from parts one and two.

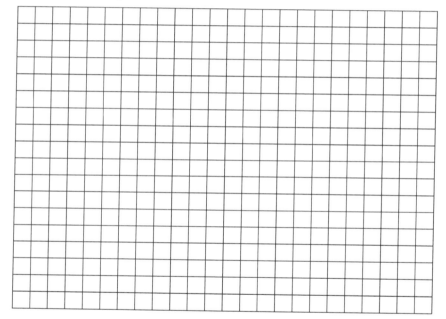

4. After how many days of visiting the pool will the special offer be a better deal? How can you explain this graphically? How might you explain this algebraically?

5. You only have $60 to spend for the summer on visiting this pool. Which offer is best for your limited budget? Explain.

Source for standard: NGA & CCSSO, 2010, p. 55.

Figure 2.7: Simultaneous linear equations task.

Visit **go.solution-tree.com/mathematicsatwork** to download a reproducible version of this figure.

First, consider whether or not this problem is a good higher-level-cognitive-demand task. Remember from chapter 1 (page 22) that good higher-level-cognitive-demand mathematical tasks provide opportunities for students to think, make use of prior knowledge, extend their thinking, and potentially develop new knowledge or solution pathways in class. The task provided in figure 2.7 assumes students know how to create equations from real-life scenarios and how to use mathematical reasoning to make real-life decisions about the information.

For this task, students will need to be able to make sense of problems and persevere in solving them (Mathematical Practice 1). The task, with its real-life scenario and the questions that call for hypothesizing, requires a certain level of attention from students and willingness to understand the meaning of the problem. If they try to work at the surface level, it will be difficult for them to engage in the task. If they model the problem and work to understand the scenario, they will be much more successful in persevering.

In this task, there is the expectation students will be able to use their prior knowledge to build functions for each scenario while also building a graphical model to represent the situation. The task provides the opportunity for students to work together (peer to peer) to ponder, hypothesize, discuss, and debate. Your role is to set up and use prompting questions to help guide students in the process of exploring the mathematical task. Examples of possible probing questions include:

- What is the mathematical task asking you to do?

- How did you build a linear function previously?

- How would you decide which is the better purchase for each swimmer? Based on what mathematical argument? Why?

- If you had $80 to spend for the summer, how would that change your response to question five, if at all?

Effective higher-level-cognitive-demand tasks provide an opportunity for students to showcase their learning in multiple ways. For example, if students worked on this particular task in small groups, each group member could take on the responsibility of developing and testing the hypothesis for each scenario presented in the problem. Some students might create a table to present their thinking, while other students might develop a graph to compare the two scenarios. How you manage students throughout the task is important to support students' opportunity to develop their proficiency with Mathematical Practice 1—"Make sense of problems and persevere in solving them."

Finally, to empower students to benefit from higher-level-cognitive-demand tasks and to subsequently develop proficiency in the practice of learning mathematics, your students need differentiated in-class support from you. You need to guide and facilitate the in-class problem-solving experience as part of a peer-to-peer, engaging student experience.

There are many actions you can take to provide differentiated support (scaffolding or advancing prompts) around meaningful mathematics tasks.

- Engage in higher-level-cognitive-demand tasks *with* students during class so they can observe effective and expected problem-solving behavior. As we will discuss in HLTA 7 (page 99), your students need constant feedback from you and each other during the learning experience the task generates.

- Encourage students to persist with a task, scaffolding as needed to support students' learning. Remember your personal classroom expectation is not needless student struggle, but rather *productive* student struggle.

- Pull from a pool of carefully selected hints or scaffolding prompts for higher-level-cognitive-demand tasks so that students can receive support to respond to a task without being given so much information that they do not need to put forth much effort (assessing questions and advancing questions).

- Help students notice the progressions of structures in the mathematics content. This will help them to better recognize types of differences and similarities between mathematical situations.

To help you plan for these mathematical task expectations, you can use the prompts in figure 2.8. Figure 2.8 provides a platform for you and your colleagues to discuss ways each team member can provide focused support for students to successfully engage in developing proficiency and perseverance with problem solving and reasoning. Engaging in discussions around the provision of differentiated support will help your team fortify itself against the temptation to *teach by telling*, particularly when students are challenged to learn the mathematics via the personal experience of a higher-level-cognitive-demand task.

Sometimes, students just need encouragement to persevere. Careful observations and engagement with your students during problem solving will help you to determine when this is needed. Carol Dweck's (2007) research provides critical insight into the nature of your praise. Students will be increasingly more apt to persevere through more challenging mathematical tasks when the nature of the praise they receive from you rewards their effort and not their ability. For example, you might say, "I like how you are using a visual model to better understand the division of fractions." This praise focuses on the student's specific effort rather than focusing on the student's ability if you say, "You are so smart! You used a model to determine the answer!"

Your team should use figure 2.8 to support an active discussion about how you not only differentiate student learning for each mathematical task used in class, but also to discuss how you personally help each student to persevere when they are stuck on the mathematical task or problems during the lesson.

Directions: Within your collaborative team, complete each statement individually. Then, collectively discuss and decide on one to two indicators for differentiated support for students to engage in higher-level-cognitive-demand mathematics tasks.

1. My lessons allow students to take risks with problem solving and reasoning (higher-level-cognitive-demand tasks) because . . .

2. Students know I have high expectations for their development and perseverance as good problem solvers because . . .

3. The types of questions that build students' proficiency, perseverance, and endurance with problem-solving and reasoning tasks are . . .

4. When students are stuck on a problem . . .

   + I do this . . .

   + They do this . . .

5. If students need a resource to model a problem-solving or reasoning task, they go to . . .

6. When I provide feedback to students during a task, my praise is mostly based on effort ("Wow, that was a great effort by your team!") or mostly based on talent ("Wow, your team must be really smart!").

Figure 2.8: Providing differentiated in-class support for higher-level-cognitive-demand tasks.

Visit **go.solution-tree.com/mathematicsatwork** to download a reproducible version of this figure.

Keeping a Sustained Focus on the Essential Learning Standards During the Unit of Instruction

Student proficiency in each Mathematical Practice through the use of higher-level-cognitive-demand tasks is a vehicle that leads to student understanding of the essential learning standard; otherwise, tasks are simply random problems for the student to solve. During instruction, you and your team can make inferences from students' engagement about the level of proficiency students are developing in select Mathematical Practices.

Consider the higher-level-cognitive-demand task (vertically stacked question parts) in figure 2.9 and the strategies for maintaining higher-level cognitive demand in figure 2.10.

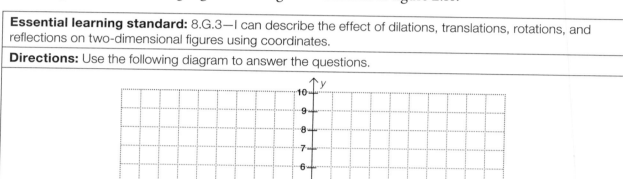

Essential learning standard: 8.G.3—I can describe the effect of dilations, translations, rotations, and reflections on two-dimensional figures using coordinates.

Directions: Use the following diagram to answer the questions.

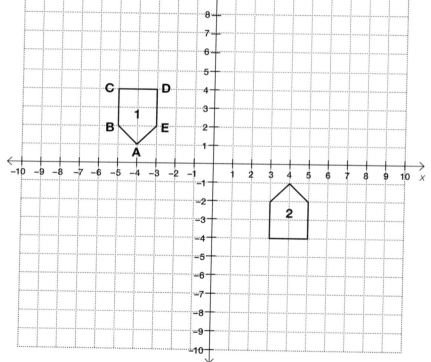

2a. Describe a series of transformations (reflection, rotation, translation, or dilation) that proves figure 1 is congruent to figure 2.

2b. Justify your answer to 2a.

2c. If you do a *translation* followed by a *reflection*, does it still map figure 1 to figure 2? How do you know?

Source: Adapted with permission from Aptakisic Junior High School, Buffalo Grove, Illinois.

Figure 2.9: Transformations using the coordinate grid.

Figure 2.10 presents strategies for maintaining a task's higher level of cognitive demand during the lesson and for making sure the students understand the essential learning standard the task addresses. Notice how this standard requires a student to describe *how they know*. Thus, you need to ensure the task fulfills the expectation of the standard. Use figure 2.10 to determine how well the task presented in figure 2.9 about the coordinate grid meets the expectations.

Strategies for maintaining higher-level cognitive demand during instruction include:

1. **Motivating students to complete the task by connecting the task to the essential learning standard for the unit**—Label the essential learning standard as an "I can" statement at the top of the task to allow students to understand the learning standard purpose for the task.

2. **Monitoring students' initial responses to the tasks or problems**—If students can easily find the solution, then the task most likely does not rise to the level of higher cognitive demand desired. Be sure to have an advancing prompt for the task to encourage deeper understanding.

3. **Ensuring tasks encourage a process of reflection and communication**—Tasks should include process moments both individually and collectively with peers.

4. **Presenting tasks in a way that requires students to apply mathematical knowledge and skills that are nonroutine**—The expectation is for students to translate their mathematical understandings from one context or representation to another as one indicator of mathematical proficiency. If students are stuck, have scaffolding prompts ready to go so they do not shut down.

5. **Expecting students to provide counterexamples and compare and contrast mathematical concepts**—Tasks shouldn't just focus on asking students to provide examples of mathematical concepts.

6. **Engaging students in solving problems orally**—Encourage the important and proper use of mathematical terminology and logical sequencing of processes.

Figure 2.10: Strategies for maintaining higher-level cognitive demand for tasks during class.

Visit **go.solution-tree.com/mathematicsatwork** to download a reproducible version of this figure.

Developing student proficiency with the Standards for Mathematical Practice and corresponding processes through the use of higher-level-cognitive-demand tasks is possible when your team has a good understanding of each Mathematical Practice, has reached agreement on what it means for students to demonstrate proficiency with each Mathematical Practice, and is committed to maintaining a level of higher cognitive demand during instruction.

Moreover, your team members benefit during the unit from the deep discussions you had before instruction began—by solving tasks together, identifying the multiple ways students might solve tasks, planning prompts to support student struggle, and planning how to address their misconceptions. Thus, how students might engage with the mathematics and demonstrate their learning *during* instruction becomes a routine part of your work together.

Your Team's Progress

As you and your team focus on using higher-level-cognitive-demand mathematical tasks effectively for student learning through demonstrations of Mathematical Practices and processes, reflect on your perspectives related to this pursuit. Your team members should individually assess the team using the status check tool in table 2.2 (page 98) to determine how well your collaborative team is currently pursuing the sixth high-leverage team action. Discuss your team's progress on using higher-level-cognitive-demand mathematical tasks effectively.

Remember that learning *how* is as important as learning *what*, and learning how manifests itself in the many decisions you make every day around the robust and intentional use of in-class formative assessment. This is the focus of the seventh high-leverage team action.

Table 2.2: During-the-Unit Status Check Tool for HLTA 6—Using Higher-Level-Cognitive-Demand Mathematical Tasks Effectively

Directions: Discuss your perception of your team's progress on the sixth high-leverage team action—using higher-level-cognitive-demand mathematical tasks effectively. Defend your reasoning.			
Stage I: Pre-Initiating	**Stage II: Initiating**	**Stage III: Developing**	**Stage IV: Sustaining**
We do not attend to or discuss Mathematical Practices or processes.	We have discussed Mathematical Practices and processes.	We consistently discuss the intent, purpose, and evidence of Mathematical Practices and processes during each unit of study.	We engage, as a team, in deep planning for Mathematical Practices and processes in our lessons.
We do not have a clear understanding of each Mathematical Practice.	We do not have a collaborative agreement on the focus of the Mathematical Practices for each unit of the course.	We teach some aspect of various Mathematical Practices as part of every daily lesson.	We use higher-level-cognitive-demand tasks as an intended activity to meet both the essential learning standards as well as Mathematical Practices and processes outlined for the lesson and the unit.
We do not use common higher-level-cognitive-demand tasks in order to develop students' Mathematical Practices.	We discuss and use some common higher-level-cognitive-demand tasks in class.	We discuss and implement collaboratively developed higher-level-cognitive-demand tasks.	We discuss and use intentional and targeted differentiated in-class supports as students engage in Mathematical Practices and processes by using our common and higher-level-cognitive-demand mathematical tasks.
We do not know the Mathematical Practices expected or demonstrated by the students assigned to other members of our team.	We have not reached team agreement on how to implement and sustain student proficiency in the Mathematical Practice expectations.	We do not collaboratively plan for Mathematical Practices, and they do not influence daily instructional plans for the unit.	We collaboratively plan for and implement Mathematical Practices and processes as part of our daily instructional plans for the unit.

Visit **go.solution-tree.com/mathematicsatwork** to download a reproducible version of this table.

HLTA 7: Using In-Class Formative Assessment Processes Effectively

[Formative assessment processes] can essentially double the speed of student learning producing large gains in students' achievement.

—W. James Popham

Recall there are four critical questions every collaborative team in a PLC asks and answers on a unit-by-unit, ongoing basis.

1. What do we want all students to know and be able to do? (The essential learning standards)

2. How will we know if they know it? (The assessment instruments and tasks teams use)

3. How will we respond if they don't know it? (Formative assessment processes for intervention)

4. How will we respond if they do know it? (Formative assessment processes for extension and enrichment)

The seventh high-leverage team action—using in-class formative assessment processes effectively—ensures you and your collaborative team reach clarity on how to effectively respond in class to the third and fourth critical questions of a PLC, How will we respond if they don't know it? How will we respond if they do know it?

High-Leverage Team Action	1. What do we want all students to know and be able to do?	2. How will we know if they know it?	3. How will we respond if they don't know it?	4. How will we respond if they do know it?
During-the-Unit Action				
HLTA 7. Using in-class formative assessment processes effectively	▢▢	▢▢	▢	▢

▢ = Fully addressed with high-leverage team action

▢▢ = Partially addressed with high-leverage team action

Through your work on this high-leverage team action, you will begin to develop student proficiency in each of the Mathematical Practices as well as other processes of student learning through engaging lessons. Descriptive feedback will also guide student actions toward achieving the essential learning standards and the Mathematical Practice focus for that lesson.

The What

The in-class, during instruction formative assessment process involves the following components:

- Unpacking the essential learning content and practice standards (see chapter 1, page 9)

- Developing and using well-designed common assessment tasks or instruments (see chapter 1, page 36)

- Collecting data through the implementation of those tasks and assessment instruments in class

- Providing clear and descriptive feedback to students

- Using the feedback (students and teachers) to adjust teaching and learning

For the feedback process to be formative, the students and the teacher must take action on the feedback.

HLTA 7 supports student proficiency in Mathematical Practices and processes through the use of effective formative assessment procedures in class. Typically, one might not think of assessment as a support for learning during class, but effective assessment (particularly the formative assessment process) is at the core of the PLC teaching-assessing-learning cycle (figure 2.1, page 80) and is critical to the learning process. It is also at the heart of NCTM's research-informed instructional practice "elicit and use evidence of student thinking." According to NCTM, "effective teaching of mathematics uses evidence of student thinking to assess progress toward mathematical understanding and to adjust instruction continually in ways that support and extend learning" (NCTM, 2014, p. 53).

According to Dylan Wiliam (2011):

> When formative assessment practices are integrated into the minute-to-minute and day-by-day classroom activities of teachers, substantial increases in student achievement—of the order of a 70 to 80% increase in the speed of learning are possible. . . . Moreover, these changes are not expensive to produce. . . . The currently available evidence suggests that there is nothing else remotely affordable that is likely to have such a large effect. (p. 161)

You and your collaborative team should not ignore this wise advice. Used effectively, this high-leverage action will have a substantial and positive impact on your students' learning. This is what Hattie (2012) means when he says, "Teacher, know thy impact" (p. ix).

During the unit of instruction, part of your team's work is to discuss how to build your confidence for using in-class formative assessment processes. This process is much more than just observing evidence of student learning (checking for understanding), which is, at best, diagnostic. For the process to also be formative, you, your students' peers, or both must provide meaningful and formative feedback during engagement with mathematics tasks and problems.

According to Reeves (2011) and Hattie (2009, 2012), there are four markers that provide a basis for effective formative feedback to your students. We have organized these markers into the acronym FAST:

1. **Feedback must be Fair:** Does your feedback rest solely on the quality of student work and not on other characteristics of the student, including some form of comparison to others in the classroom?

2. **Feedback must be Accurate:** Is the feedback during the in-class activity actually correct? Do students receive prompts, solution pathway suggestions, and discourse that are effective for understanding the mathematical task or activity, as you tour the room observing individuals and small-group teams of students as they work and check for understanding?

3. **Feedback must be Specific:** Does the verbal feedback students receive contain enough specificity to help them persevere and stay engaged in the mathematical task or activity process? Does the feedback help students to get "unstuck" or to advance their thinking as needed? (For example, "Work harder on the problem" is not helpful feedback for a student.)

4. **Feedback must be Timely:** As you tour the room and listen in on the peer-to-peer conversations, is your feedback immediate and corrective to keep students on track for the solution pathway?

However, FAST meaningful feedback from you and by student peers alone is not sufficient for student learning. If, during the best teacher-designed moments of classroom formative assessment processes, teachers fail to support students *taking action* on areas of difficulty (step three of the PLC teaching-assessing-learning cycle, page 80), then the cycle of learning, assessing, and continued learning stops for the student (Wiliam, 2011).

The How

According to W. James Popham (2011), when teachers use formative assessment well:

> It can essentially double the speed of student learning producing large gains in students' achievement; and at the same time, it is sufficiently robust so different teachers can use it in diverse ways and still get great results with their students. (p. 36)

So what does this have to do with the work of your collaborative team? Your work should begin with making sure your team is clear about the often used (and misused) in-class teaching technique of checking for understanding.

Checking for Understanding Versus Using Formative Assessment Processes

You and your colleagues need to be clear about the difference between checking for understanding in class (which has minimal impact on student learning) and using formative assessment processes (which includes the classroom elements of formative feedback and student action on feedback). Think back to table 2.1 (page 81) and your responses to the definitions of these concepts. To facilitate your discussion of these differences, engage with your team in the formative assessment process tool in figure 2.11 (page 102).

Directions: Answer the following questions.

1. How do you currently check for student understanding on a daily basis? Write your responses in the Checking for Understanding column. For each response, indicate whether you use whole-group discourse (teaching at the front of the room) with a *W*, small-group discourse (students working with their peers) with an *S*, or independent practice (students working individually) with an *IP*.

2. What do you and your colleagues believe is the difference between solely checking for understanding and using formative assessment processes in the classroom during instruction?

3. How can you implement formative assessment processes in the classroom each day?

Remember, for the process to be formative, students must actually take action on the feedback they receive. In the Using Formative Assessment Processes column, explain ways that you could improve each check for understanding you listed to become a moment of formative assessment for your students.

Checking for Understanding	Using Formative Assessment Processes

Figure 2.11: Checking for understanding versus using the formative assessment process tool.

Visit **go.solution-tree.com/mathematicsatwork** to download a reproducible version of this figure.

As you learn more about how to engage in this high-leverage teacher and team action you can complete the figure further. You promote conceptual understanding when you explicitly make, or ask students to make, connections among ideas, facts, and procedures (Hiebert & Grouws, 2007). The following five actions help students make these connections (Kanold, Briars, & Fennell, 2012).

1. Challenging students to think, reason, and make sense of what they are doing to solve mathematics problems

2. Posing whole-group and small-group questions that stimulate students' thinking and reasoning and asking them to justify their conclusions, solution strategies, and processes

3. Expecting students to evaluate and explain the work of other students and engage in peer-to-peer discourse by comparing and contrasting different solution pathways for the same problem, while receiving formative feedback on the accuracy of those pathways

4. Expecting students to engage in productive struggle and action necessary to find successful solutions to the tasks

5. Asking students to represent the same ideas in multiple ways—using multiple representations, such as numerical tables, graphs, or models to demonstrate understanding of a concept

What does a robust formative assessment process look like in the classroom? Consider figure 2.12 (page 104). In figure 2.12, you and your team examine a sixth-grade task: understanding and using plotting of points on a coordinate grid. In this example that uses a real-life context, you examine key formative assessment process questions to consider as part of the task's formative process. You and your team should take the time together to respond to the questions in figure 2.12.

At the heart of a robust vision for mathematics instruction is what your students are *doing* during class. How are they engaged? With whom are the most noteworthy conversations taking place in your classroom: student-to-student or teacher-to-student? Moreover, in your classroom, do your students see each other as reliable and valuable resources for learning?

The Mathematical Practices and processes, and good mathematics instruction in general, are best implemented via small-group discourse, providing opportunities for both teachers and student peers to offer meaningful formative feedback to each other during class. For additional information on implementing mathematical discourse in the classroom, see the research-informed instructional practices "facilitate meaningful mathematical discourse" and "pose purposeful questions" in *Principles to Actions* (NCTM, 2014, pp. 29–41). However, for feedback to be effective, students must take action on the formative feedback they receive.

Directions: Read the following scenario, and respond to the formative assessment process questions.

Mrs. Leibach recalls that the last time she used the following problem task in a lesson, she needed more preparation for determining what students were or were not learning as they engaged in the problem-solving task. This time, during instruction, she was committed to gathering student-learning data so she could make better decisions about the type of formative feedback to provide and the next steps in instruction and student action to increase the potential for student learning.

Riding to the Park Task: You are riding to the park. You start off at your house, located at A (–5, 2). You ride to the store located at B (–5, –8) to buy some sunglasses. After that, you pick up your friend at C (1, –8). Together, you ride to the park located at D (1, 9). If one unit on the coordinate grid equals 1 block, how many blocks did you ride? If you rode from your house straight to the park, how many blocks would you ride? Explain your thinking and the process you used to solve the problem.

1. Mrs. Leibach wanted a series of questions ready to ask her students in order to probe their thinking. Two question starters appear below. What are two more question starters she could use?

   + What do you know about the coordinate grid? How can you use that to solve this problem?

   + Where do you start with the information given in this problem? Why?

   If students need more prompting, Mrs. Leibach could ask the students to plot the points listed, and then see if that helps them answer her questions.

2. What questions do you anticipate the students will ask Mrs. Leibach?

3. With which problem-solving strategies (such as working backward) might Mrs. Leibach observe the students engaged?

4. Because student responses to the task will vary, how do you propose Mrs. Leibach facilitate the process?

5. What misconceptions do you anticipate Mrs. Leibach will uncover? What suggestions do you have to help her address these misconceptions?

Figure 2.12: Mathematics instruction scenario.

Visit **go.solution-tree.com/mathematicsatwork** to download a reproducible version of this figure.

Going beyond checking for understanding from the front of the classroom and moving into using meaningful formative feedback as part of instruction require rich mathematical tasks that support robust student discourse and well-managed activities as you orchestrate advancing and assessing prompts—allowing student-engaged exploration, discussions, and perseverance with peers.

Work in your collaborative team with figure 2.13 (pages 105–106) to review the tasks that most closely relate to the grade-level responsibilities of your collaborative team. You can use figure 2.11, the checking for understanding versus using the formative assessment process tool (page 102), to help you determine formative assessment prompts that you could use for each task listed in figure 2.13—specifically, what questions might you ask of students to keep them engaged during the lesson?

Grade 6 Essential Learning Standard 6.NS.5: Understand that positive and negative numbers are used together to describe quantities having opposite directions or values (e.g., temperature above/below zero, elevation above/below sea level, credits/debits, positive/negative electric charge); use positive and negative numbers to represent quantities in real-world contexts, explaining the meaning of 0 in each situation.

Jamal is filling bags with sand. All of the bags are the same size. Each bag must weigh less than 50 pounds. One sand bag weighs 57 pounds, and another sand bag weighs 41 pounds. Explain whether Jamal can put sand from one bag into the other so that the weight of each bag is less than 50 pounds.

Grade 7 Essential Learning Standard 7.RP.1: Compute unit rates associated with ratios of fractions, including ratios of lengths, areas, and other quantities measured in like or different units. For example, if a person walks ½ mile in each ¼ hour, compute the unit rate as the complex fraction ½/¼ miles per hour, equivalently 2 miles per hour.

Mr. Ruiz is starting a marching band at this school. He first does research and finds the following data about other local marching bands.

	Band 1	Band 2	Band 3
Number of Brass Instrument Players	123	42	150
Number of Percussion Instrument Players	41	14	50

Part A

Type your answer in the box. Backspace to erase.

Mr. Ruiz realizes that there are [＿＿＿] brass instrument player(s) per percussion player.

Part B

Mr. Ruiz has 210 students who are interested in joining the marching band. He decides to have 80% of the band be made up of percussion and brass instruments. Use the unit rate you found in Part A to determine how many students should play brass instruments.

Show or explain all your steps.

Figure 2.13: Grade-level tasks revisited for formative assessment feedback and action considerations.

continued →

Grade 8 Essential Learning Standard 8.EE.7: Solve linear equations in one variable.

Consider this equation:

$c = ax - bx$

Joseph claims that if a, b, and c are non-negative integers, then the equation has exactly one solution for x. Select all cases that show Joseph's claim is incorrect.

❏ $a - b = 1, c = 0$

❏ $a = b, c \neq 0$

❏ $a = b, c = 0$

❏ $a - b = 1, c \neq 1$

❏ $a \neq b, c = 0$

Source for the task: Smarter Balanced Assessment Consortium, n.d. Used with permission.
Source for standards: NGA & CCSSO, 2010, pp. 42, 48, 54.

Figure 2.13: Grade-level tasks revisited for formative assessment feedback and action considerations (continued).

Visit **go.solution-tree.com/mathematicsatwork** to download a reproducible version of this figure.

In general, as you plan each lesson, you should ask yourself about the particular student solution strategies or pathways you should observe, provide scaffolding or advancing feedback for, and expect action on, as students engage in the given mathematical tasks used to unfold any lesson.

Developing the In-Class Formative Assessment Process Through Student Reasoning

Consider the Mathematical Practice—"Reason abstractly and quantitatively." One of the primary outcomes of middle school mathematics standards is that your students will experience various levels of reasoning every day in the classroom. Your collaborative team should prepare and plan for how your in-class formative assessment processes will support student reasoning every day as you use higher-level-cognitive-demand tasks in class.

What does it mean to reason? Do students reason alone? With their teacher? With each other? Out loud? In debate? On paper or with technology? The answer is yes to all of these methods of demonstrating reasoning. The higher the cognitive demand of the mathematical tasks you present for students to do in class—not for them to watch *you* do—the greater the levels of complexity of the reasoning you display. Your level of commitment to ensuring that reasoning takes place in your classroom every day will determine the kinds of mathematical experiences you provide students and how you assess their understanding in the contexts of these experiences. Consider the higher-level-cognitive-demand reasoning task in figure 2.14.

Directions: Use the graph of the continuous function to answer the following questions.

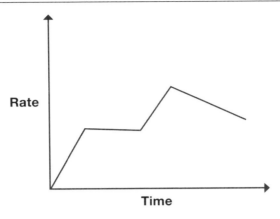

1. Explain how you might interpret this graph. As time (*x*) increases, what describes the nature of the changes in the rate (*y*)?

2. Create a real-life scenario to describe the rate of change represented by the graph. Be sure to include details to represent each moment in time the rate changes.

3. **Extension:** Describe the nature of both the direction and the speed of the path if someone were to walk at the rate of change shown in the graph.

Figure 2.14: Grade 8 higher-level-cognitive-demand reasoning task.

Visit **go.solution-tree.com/mathematicsatwork** to download a reproducible version of this figure.

Do this task with your team, and determine the type of reasoning and various solution pathways you might expect students to pursue when engaging with this task in class. All students would be expected to do the task as shown. More advanced students could also reason through an extension to the task. You can use the rubric in figure 2.15 to assess student progress (and proficiency) in general for Mathematical Practice 2, "Reason abstractly and quantitatively," and, more specifically, for reasoning through the sample mathematical task in figure 2.14.

Falls Far Below Learning Standard	Approaches Learning Standard	Meets Learning Standard	Exceeds Learning Standard
1	2	3	4
The answer is missing or incorrect, and there is little evidence of reasoning or application of mathematics to solve the problem.	The answer is missing or incorrect, but there is evidence of some reasoning or application of mathematics to solve the problem.	The answer is correct with some evidence of reasoning or application of mathematics to solve the problem, *or* the answer is incorrect, but there is substantial evidence of solid reasoning or application of mathematics to solve the problem.	The answer is correct, and substantial evidence of solid reasoning or application of mathematics to solve the problem supports it.

Source: Adapted from EngageNY, 2013.

Figure 2.15: Demonstrating reasoning rubric.

In using the higher-level-cognitive-demand reasoning task in figure 2.14 during instruction, your collaborative team decides how to use the task to facilitate student learning of essential learning standard 8.F.4, "Interpret the rate of change and initial value of a linear function in terms of the situation it models, and in terms of its graph or a table of values" (NGA & CCSSO, 2010, p. 55). In and of itself, the higher-level-cognitive-demand task does not guarantee student reasoning and understanding will take place, however:

> Accomplished teachers deliberately structure opportunities for students to use and develop appropriate mathematical discourse as they reason and solve problems. These teachers give students opportunities to talk with one another, work together in solving problems, and use both written and oral discourse to describe and discuss their mathematical thinking and understanding.
>
> As students talk and write about mathematics—as they explain their thinking—they deepen their mathematical understanding in powerful ways that can enhance their ability to use the strategies and thought processes gained through the study of mathematics to deal with life issues. (National Board for Professional Teaching Standards, 2010, p. 57)

The manner in which you facilitate discourse is critical when creating and supporting a classroom learning environment that values reasoning and sense making. You decide what thinking to share and whose voices are heard, which has a profound impact on how you create knowledge as well as who plays a role in that creation. "Teachers, through the ways in which they orchestrate discourse, convey messages about whose knowledge and ways of thinking and knowing are valued, who is considered able to contribute, and who has status in the group" (NCTM, 1991, p. 20).

Small-group discourse with a student team can also help you meet the expectations of a Mathematical Practice such as, "Construct viable arguments and critique the reasoning of others" and supports the research-informed instructional practice "facilitate meaningful mathematical discourse" in *Principles to Actions* (NCTM, 2014). Discourse among small groups of students opens a window into student reasoning about mathematics; you hear and see the connections students make and the questions they ask, as well as the obstacles or misconceptions that can hinder their conceptual understanding. You are then in a better position to make decisions and implement strategies to support learning and extend the level of reasoning and problem solving expected for the Mathematical Practice, "Make sense of problems and persevere in solving them." Small-group student dialogue creates a community of learners who collectively build mathematical knowledge and proficiencies *together*.

You can use the walk-around tool in figure 2.16 (page 110) to collect information and provide feedback on what you observe when student teams are engaged in the mathematical tasks and activities you plan for class. As part of this formative process, you provide guidance and scaffold questions to support student learning and perseverance on the task. You also determine if you need to make adjustments in your whole-group instruction to develop students' reasoning skills. If a student team is ready, encourage the students to try the task's extension activity as well.

You and your collaborative team should discuss how a tool like this might be useful in providing information about students' development of proficiency to reason abstractly and quantitatively during your lessons.

You support student proficiency with *all* Mathematical Practices when you make your plans to implement effective formative assessment strategies explicit.

Student Teams (List Team Member Names)	Level 1 — The student team is working the task but is constantly stuck. Students do not generally take corrective action on the teacher's feedback and scaffolding prompts.	Level 2 — The student team is working the task by connecting to prior knowledge, engaging in conversations, and taking corrective action to the teacher's scaffolding prompts.	Level 3 — The student team is working the task by engaging in accurate sense making and reasoning and using multiple connections with minimal teacher feedback.	Level 4 — The student team has worked the task and completed it correctly and is engaging in an extension to the problem (advancing prompts) provided by the teacher.
Student team 1				
Student team 2				
Student team 3				
Student team 4				
Student team 5				
Student team 6				
Student team 7				
Student team 8				

Overall observations of student reasoning and engagement in the higher-level-cognitive-demand task:

Figure 2.16: Walk-around formative assessment tool.

Visit **go.solution-tree.com/mathematicsatwork** to download a reproducible version of this figure.

Planning for the Formative Assessment Process

You and your collaborative team can use the tools in this section—the example tasks from grade 6, grade 7, and grade 8 (figures 2.17, 2.18, and 2.19, respectively, on pages 112, 113, and 114)—to explore elements of planning for the formative assessment process. As you use higher-level-cognitive-demand tasks within the context of the Mathematical Practices and processes (what students are to *do*), your plans for creating a formative assessment process in class can unfold. Keep in mind that the practices focus on student engagement with tasks, rather than on teacher engagement. The key is how you choose to orchestrate and effectively provide feedback to students as they work on the mathematical task.

Remember, the best structural vehicle for you to lead the formative process in class is to observe for student understanding, provide feedback to students based on your diagnosis, and help students take action on that feedback during your use of small-group student discourse. Work on solutions to the tasks in figures 2.17, 2.18, and 2.19, and then answer the questions for how you would use each task to develop student reasoning and productive engagement in the task during class. Visit **go.solution-tree .com/mathematicsatwork** for blank templates for figures 2.17, 2.18, and 2.19 that can be used with any task you choose.

Grade: 6

Essential learning standard: 6.NS—I can describe a set of rational numbers. I can explain the meaning of zero in real-world situations.

1. Which number is closest to 2 on a number line, 5 or –2? Show your solution by using the number line. Explain your reasoning.

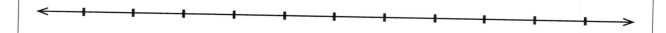

2. Using what you know about positive and negative numbers, create a situation that would result in zero as a solution. Use a number line to justify your example choice.

Which Mathematical Practice can students best develop proficiency in by working on this task? Why?	What types of scaffolding questions can you ask students to help guide their work on this task?	What can you learn about the mathematics that students know when they work on this task?	What can you learn about the mathematics that challenges students when they work on this task?
How do you plan to provide feedback to student solution pathways and explanations?	What changes will you make to the task the next time you use it in instruction?	What type of student responses demonstrate deep understanding as a result of engaging in this task?	How will you ensure all students take action on your feedback during the task?

Source: Tasks adapted from Schaumburg School District 54, Schaumburg, Illinois.
Source for standard: NGA & CCSSO, 2010, p. 42.

Figure 2.17: Planning for the formative assessment process example—grade 6.

Visit **go.solution-tree.com/mathematicsatwork** to download a reproducible version of this figure.

Grade: 7	

Essential learning standards: 7.NS and 7.EE—I can describe how quantities represented in different forms within an expression are related.

1. In order to determine the product (9.7)(–2), Alan decided to try (10 – 0.3)(–2). Is this an equivalent expression?

   a. Justify why this will or will not work.

   b. What is another equivalent expression he could have used in order to evaluate 9.7(–2)?

2. Given a square fenced yard, as shown in the picture, write four different numerical expressions to find the total number of tiles in the border. Show how each of the expressions relates to the diagram and demonstrate that the expressions are equivalent. Which expression do you think is most useful? Justify your reasoning.

   Numerical Expressions

   a.

   b.

   c.

   d.

Which Mathematical Practice can students best develop proficiency in by working on this task? Why?	What types of scaffolding questions can you ask students to help guide their work on this task?	What can you learn about the mathematics that students know when they work on this task?	What can you learn about the mathematics that challenges students when they work on this task?
How do you plan to provide feedback to student solution pathways and explanations?	What changes will you make to the task the next time you use it in instruction?	What type of student responses demonstrate deep understanding as a result of engaging in this task?	How will you ensure all students take action on your feedback during the task?

Source: Tasks adapted from Schaumburg School District 54, Schaumburg, Illinois.
Source for standards: NGA & CCSSO, 2010, pp. 48, 49.

Figure 2.18: Planning for the formative assessment process example—grade 7.

Visit **go.solution-tree.com/mathematicsatwork** to download a reproducible version of this figure.

Grade: 8

Essential learning standards: 8.G.2 and 8.G.4—I can verify that two-dimensional figures are congruent using transformations and demonstrate the similarity of two-dimensional figures using transformations.

State whether each of the following sets of figures are similar, congruent, or neither. If they are similar or congruent, explain the transformation required to create the image from A to B. If they are neither, explain why you think so.

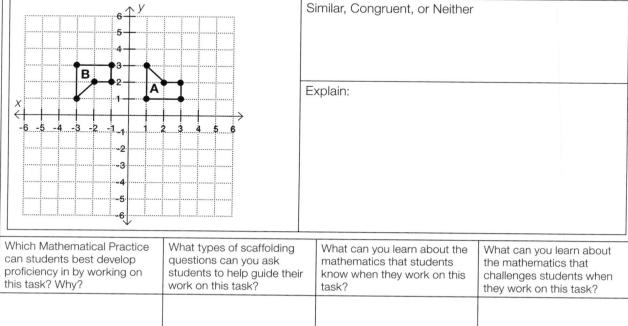

Similar, Congruent, or Neither

Explain:

Which Mathematical Practice can students best develop proficiency in by working on this task? Why?	What types of scaffolding questions can you ask students to help guide their work on this task?	What can you learn about the mathematics that students know when they work on this task?	What can you learn about the mathematics that challenges students when they work on this task?
How do you plan to provide feedback to student solution pathways and explanations?	What changes will you make to the task the next time you use it in instruction?	What type of student responses demonstrate deep understanding as a result of engaging in this task?	How will you ensure all students take action on your feedback during the task?

Source: Tasks adapted from Schaumburg School District 54, Schaumburg, Illinois.
Source for standards: NGA & CCSSO, 2010, pp. 55, 56.

Figure 2.19: Planning for the formative assessment process example—grade 8.

Visit **go.solution-tree.com/mathematicsatwork** to download a reproducible version of this figure.

Do your students gradually take more responsibility for their learning by reflecting on their work and viewing solution mistakes as learning opportunities? This is essential for student learning to become formative in your classroom. As Dylan Wiliam (2007) indicates, in order to "improve the quality of learning within the system, to be formative, feedback needs to contain an implicit or explicit recipe for future action" (p. 1062).

Students make errors, receive feedback on those errors, and take action to correct their mistakes in reasoning. Students then view assessment of their work as something they *do* in order to focus their energy and effort for future learning. A great place for this type of student reflection is during small-group discourse as your students work together in teams on various problems or tasks and you walk around the room, check for understanding, and provide students with meaningful feedback and scaffolding and advancing prompts that differentiate the task for them.

Thus, students and teachers share the responsibility to successfully implement the in-class formative assessment processes and practices. When your students can demonstrate understanding through reasoning mathematically, they connect better to the essential learning standard for the unit and can reflect on their individual progress toward that learning target. You support students' progress by using immediate and effective feedback during the daily classroom conversations, whether in whole-group or small-group moments.

You can apply formative assessment processes to written work students complete, but it shouldn't be limited to pencil-and-paper submissions like exit slips and quizzes, which tend to only be diagnostic checks for understanding and fall far short of the feedback and action benefits of formative assessment on student learning. Figure 2.20 (page 116) represents different ways in which formative assessment can occur in a classroom.

Formative Assessment	Description
Written feedback	When students are given a task, the teacher provides specific written comments to guide student thinking and action.
Teacher observation while walking around	The teacher observes and listens to student conversations and uses questioning prompts and strategies to push student thinking or support student thinking.
Questioning—Whole group or small group	Teacher may ask questions such as the following: "Why?" "Could you explain it another way?" "How does this connect with _____?"
Exit slips	The teacher uses a specific question or task that ties to content from the class. (Keep higher-level cognitive demand on these questions). However, for the exit slip to be formative, there must be student action on the feedback provided.
Voting	Individual students or small groups of students vote to answer a question the teacher poses. Voting can involve such tools as: + Whiteboards + Clickers with interactive whiteboards + Websites like Poll Everywhere (www.polleverywhere.com) or Infuse Learning (www.infuselearning.com) In the aftermath of the vote, students must take action and make corrections as needed.
Peer feedback through group work	When given higher-level-cognitive-demand tasks that require collaboration with a group, students can use each other for feedback, support, and in-time action.
Voting with your feet	The teacher poses agree or disagree questions. Students move to either side of the room to indicate whether they agree or disagree, and then they self-correct as needed.
Nonverbal check	Using a scale of 1–5, students hold up the number of fingers to their chest to indicate their comfort level with the content. Using thumbs-up and thumbs-down turns this strategy into a formative learning activity based on immediate response to the vote.
Quizzes	When using unit quizzes or homework quizzes, there must be student follow-up action based on the results to make this formative. A quiz retake process helps support student action on the data with teacher feedback for each standard.
A learning management system (like Google Forms, Schoology, Edmodo, and so on)	Students take online quizzes and get their results immediately, and the teacher can see all student results. The teacher can use this for a quick student regroup to then re-engage students in specific areas of weakness. Students would also need to take action to make corrections.
Online discussion forums (like Schoology, Edmodo, Socrative, TodaysMeet, and so on)	Students participate in online classroom discussions where they share their thinking, read classmate explanations, and learn from each other.

Figure 2.20: Sample formative assessment strategies for your classroom.

Visit **go.solution-tree.com/mathematicsatwork** to download a reproducible version of this figure.

Managing Students in Peer-to-Peer Productive Discourse

Managing students in peer-to-peer productive discourse is critical for effective implementation of lower- and higher-level-cognitive-demand tasks. When you set your student teams to work, do they know the expectation for teamwork, or do they rely on one or two students to get them started? Your students will need structure to support meaningful mathematics discourse, engagement, and action.

As students are engaged in the mathematics, structures for their engagement help to set the expectations for teamwork. Students should know the expectations for team interactions and their rights and responsibilities to their team. Your collaborative team can create posters to share these expectations with students, such as the one shown in figure 2.21.

Working in Teams

Rights	Responsibility
Each student has the right to:	Each student should:
• Have a voice in discussions	• Actively listen to all team members
• Ask for assistance	• Help others when asked
• Embrace mistakes	• Provide positive feedback
• Express his or her opinion	• Be encouraging and open to other perspectives
• Learn from his or her team	• Respect others' rights
• Disagree with respect	• Seek consensus
• Learn rich mathematics	• Ensure success for all team members

Source: Adapted from South Mountain High School, Phoenix, Arizona, Mathematics Department.

Figure 2.21: Classroom discourse rules and responsibilities.

Once students understand the expectations for team behavior, you can incorporate structure into rich mathematical tasks so all members have a role. For example, there are several cooperative learning structures you can use to encourage participation, the key to engaging students in mathematical thinking (Johnson & Johnson, 1999; Johnson, Johnson, & Holubec, 2008; Kagan, 1994; Kagan & Kagan, 2009).

Small-group discourse in the classroom should have the following five key elements (Kagan, 1994; Kagan & Kagan, 2009).

1. **Positive interdependence:** Structure the task so group members must work together to complete the task with each team member having a unique contribution to the team. Team members will rely on each other to complete the task.

2. **Individual accountability:** Make all students accountable for completing the task.

3. **Face-to-face feedback and interactions:** Students should provide feedback to each other, negotiating solution strategies, taking risks, testing conjectures, and ensuring that everyone on the team understands the team's thinking.

4. **Interpersonal and small-group relational skills:** Team members use appropriate skills to support each other, develop trust, and build a community of learners.

5. **Team processing:** Students self-assess how they are working together, and teachers provide feedback to teams on their ability to work cooperatively.

As you and your collaborative teams plan for student learning experiences, use the tool in figure 2.22 to list how the activity promotes the five key elements of effective small-group student discourse.

Essential learning standard: _____
Small-group discourse learning activity:
Directions: How does this learning activity promote each of the following discourse criteria for each student team?

Positive Interdependence	Individual Accountability	Face-to-Face Feedback and Interactions	Interpersonal and Small-Group Relational Skills	Team Processing

Figure 2.22: Small-group discourse peer-to-peer engagement tool.

Visit **go.solution-tree.com/mathematicsatwork** to download a reproducible version of this figure.

For more explicit information on how to manage peer-to-peer small-group discourse in your class, see *Five Practices for Orchestrating Productive Mathematics Discussions* (Smith & Stein, 2011).

Connecting In-Class Formative Assessment Processes to the End-of-Unit Common Assessment

One outcome of effective formative assessment is that students will be prepared to showcase their learning on the common end-of-unit assessment you prepared before the unit began. Recall the end-of-unit grade 8 functions test in figure 1.25 (pages 59–63).

The expectation is that before students engage in this common end-of-unit assessment, they have received timely and specific feedback from formative assessment opportunities in class. Through collaborative teamwork, you and your team members can benefit from the data you collect during the in-class formative assessment process. These actions should result in improved student performance on the end-of-unit assessment.

So, what should happen during the unit instruction that will support students' success on this common end-of-unit assessment? You can use the tool in figure 2.23 (page 120) to evaluate your current progress toward connecting your in-class formative assessment process work to the end-of-unit assessment.

Formative Assessment and Your Tier 1 Differentiated Response to Learning

The recommendations of this high-leverage team action are linked closely to your Tier 1 RTI responses in class. These interventions are the core of an RTI model designed to support learning in class by every student. Douglas Fisher, Nancy Frey, and Carol Rothenberg (2010) suggest that "interventions are an element of good teaching" (p. 2), and these interventions begin in the classroom.

Tier 1 interventions are your first line of defense for struggling students and include your *differentiated* response to learning as you provide students and student teams with scaffolding prompts to help them think of other ways to solve a problem or higher-level-cognitive-demand task because you know they can explore a concept more deeply from a different point of view.

Within the RTI framework, differentiation is explicit and is purposefully planned by your team. You and your collaborative team can use formative assessment data, knowledge of students' prior knowledge, language, and diverse culture to offer students in the same class different teaching and learning opportunities to address student learning needs—especially as you give students feedback on their progress and they take action with their peers. By using formative assessment strategies you are not making the content easier; you are making the content more accessible by the time the class period ends.

Your willingness and adaptability to try different teaching strategies and solution pathways to help all students understand the learning of the standard are in and of themselves Tier 1 intervention responses. No matter how great the lesson, if it doesn't result in student demonstrations of understanding, then seeking other ways to teach the standard is a great Tier 1 teacher-led response. For more guidance on the nature of your Tier 1 response, see chapter 5 in *Common Core Mathematics in a PLC at Work, Grades 6–8* (Kanold et al., 2013).

Directions: Select your next upcoming common assessment. Reflect on each statement individually, and then discuss several actions taking place during instruction that will prepare students for success on that common assessment instrument.

	1 Strongly Disagree	2	3 Agree	4	5 Strongly Agree
1. Teachers on our team use a variety of formative assessment strategies to encourage students to use and persevere through both standard and invented strategies.					
2. Our students represent their learning in multiple ways—such as with tables, graphs, functions, equations, and models.					
3. Teachers on our team demonstrate that the answer to a mathematics problem or task is not all that matters in learning; reasoning and explaining why and how are also important.					
4. Our students improve proficiency with both concepts and procedures through experiences with higher- and lower-level-cognitive-demand tasks.					
5. Our students increase their problem-solving and reasoning skills by solving problems together, so teachers present students with higher-level-cognitive-demand tasks without lowering the demand during instruction and then require students to justify their responses and reasoning for the solution pathways to such tasks.					

Discussion based on your evaluation for each of the five statements:

Figure 2.23: Connecting in-class formative assessment process work to the end-of-unit assessment tool.

Visit **go.solution-tree.com/mathematicsatwork** to download a reproducible version of this figure.

Your Team's Progress

It is helpful to diagnose your team's current reality and action during the unit. Ask each team member to individually assess your team on the seventh high-leverage team action using the status check tool in table 2.3. Discuss your perception of your team's progress on using in-class formative assessment processes effectively. As your team seeks stage IV—sustaining—you will increase rigor, coherence, and fidelity toward student improvement and demonstration of proficiency in various Mathematical Practices and processes during the unit of instruction. You will also support deeper levels of student understanding for the learning standards, increasing students' chances for success on the end-of-unit assessment.

Table 2.3: During-the-Unit Status Check Tool for HLTA 7—Using In-Class Formative Assessment Processes Effectively

Directions: Discuss your perception of your team's progress on the seventh high-leverage team action—using in-class formative assessment processes effectively.			
Stage I: Pre-Initiating	**Stage II: Initiating**	**Stage III: Developing**	**Stage IV: Sustaining**
We do not attend to or discuss formative assessment processes in our instruction.	We have discussed formative assessment processes and do not need to do so again.	We emphasize to students the value of using formative assessment feedback during class and taking action during class.	We always use in-class formative assessment processes to inform students' reasoning and learning.
We do not have a clear understanding of the difference between checking for understanding and formative assessment processes.	We discuss and use checking for understanding methods in class, but do not provide feedback or expect action.	We plan for student reasoning and sense making, and it is built into a formative feedback process in class.	We engage, as a team, in deep planning for formative assessment processes through the use of small-group discourse with teacher feedback and student action.
We do not know the in-class formative assessment methods used or expected by other members of our team.	We do not plan for how to engage students in formative assessment processes and practices as part of our team focus for each unit.	We plan for effective small-group student team activities as a way to promote the formative assessment process in class.	We discuss and implement well-developed formative assessment procedures to use with many of our higher-level-cognitive-demand tasks.
We do not use in-class formative feedback with student action in order to develop students' mathematical practice and learning.	We have not yet reached team agreement on how to use differentiated and targeted in-class Tier 1 RTI supports as part of our instruction.	We do not know the Tier 1 RTI response in class by other members of our team.	We discuss and use intentional differentiated and targeted in-class Tier 1 RTI supports as students engage in common higher-level-cognitive-demand mathematical tasks.

Visit **go.solution-tree.com/mathematicsatwork** to download a reproducible version of this table.

So much of student learning depends on the decisions you make every day. Will you use a robust set of mathematical tasks? Will you become intentional about using effective in-class formative assessment processes that exhibit meaningful feedback with subsequent student action? It is a lot to ask as part of your planning and teaching every day.

To help you become more organized in how you might think through your daily mathematics instruction, this chapter concludes with the eighth high-leverage team action—using a lesson-design process for lesson planning and collective team inquiry.

You can ensure student learning of the unit content standards and proficiency in Mathematical Practices and processes when your collaborative team has a good understanding of each Mathematical Practice, agrees on the meaning of *proficiency* for each Mathematical Practice, and uses effective formative assessment processes during instruction. All of these elements come together in this final section of chapter 2.

HLTA 8: Using a Lesson-Design Process for Lesson Planning and Collective Team Inquiry

Visible learning means an enhanced role for teachers, as they become evaluators of their own and each other's teaching.

—John Hattie

This eighth high-leverage team action highlights the team's work to develop and use daily mathematics lessons that encourage students to think deeply about the learning standards and to demonstrate proficiency in the Mathematical Practices. Planning several well-designed mathematics lessons during the unit brings together all of the elements of the seven high-leverage team actions and becomes part of your personal and ongoing team challenge as your teaching becomes more visible to one another.

Recall there are four critical questions every collaborative team in a PLC asks and answers on a unit-by-unit, ongoing basis.

1. What do we want all students to know and be able to do? (The essential learning standards)

2. How will we know if they know it? (The assessment instruments and tasks teams use)

3. How will we respond if they don't know it? (Formative assessment processes for intervention)

4. How will we respond if they do know it? (Formative assessment processes for extension and enrichment)

High-leverage team action 8—using a lesson-design process for lesson planning and collective team inquiry—ensures your team reaches clarity on all four of the PLC critical questions within your mathematics lesson. The lesson-design process ensures all team members possess a strong understanding of the intent of the standard for the lesson, the purpose of the tasks to be used, and the expectation of student mastery of the standard.

High-Leverage Team Action	1. What do we want all students to know and be able to do?	2. How will we know if they know it?	3. How will we respond if they don't know it?	4. How will we respond if they do know it?
During-the-Unit Action				
HLTA 8. Using a lesson-design process for lesson planning and collective team inquiry	▢	▢	▢	▢

▢ = Fully addressed with high-leverage team action

The What

Effective mathematics instruction rests, in part, on your careful planning (Morris, Hiebert, & Spitzer, 2009), as well as on the careful planning of your team. Your collaborative teams are uniquely structured to provide time and support needed to interpret the essential learning standards, embed Mathematical

Practices and processes into daily lessons, and reflect together on the effectiveness of your lesson implementation. The professionalism principle in *Principles to Actions* (NCTM, 2014) specifically recommends that teachers collaboratively develop common instructional plans and reflect on their effectiveness (together).

In this eighth high-leverage team action, you can use the lesson-planning tool during unit instruction. This tool (see figure 2.24, pages 125–126), adapted from the *Common Core Mathematics in a PLC at Work* series, is a way to support the focus and design of your mathematics tasks. It can help you plan for formative assessment feedback questions and evidence *during* the mathematics lesson. Consider the following three questions.

1. How do we expect students will express their ideas, questions, insights, and difficulties?

2. Where and when will and should the most significant conversations take place (student to teacher, student to student, or teacher to student)?

3. How approachable and encouraging should we be as students explore? Do students use and value each other as reliable and valuable learning resources?

Your collaborative team is most likely using some type of lesson-planning format or tool. What separates the Mathematical Practices lesson-planning tool from most other lesson-planning models is its focus on developing students' proficiency with Mathematical Practices *and* content standards for the lesson.

This lesson-planning tool also specifically asks you and your team to consider both student and teacher actions during lesson planning, which helps you see learning through students' eyes and helps students become their own teachers. Thus, learning becomes more visible to you, your colleagues, and your students. One particularly powerful collaborative tool is lesson study. Lesson study is very effective (Gersten, Taylor, Keys, Rolfhus, & Newman-Gonchar, 2014; Hiebert & Stigler, 2000) as a collaborative protocol with a high impact on teacher professional learning. Your team can also use the Mathematical Practices lesson-planning tool as you participate in collective inquiry during the unit. In addition, the tool emphasizes the importance of selecting and using higher-level-cognitive-demand tasks along with the more common lower-level-cognitive-demand tasks to support students' learning of mathematics.

The planning tool also supports your focused response to the critical beginning-of-class routines and the end-of-class routines for summary. Your opening and closing class activities have a significant emotional impact on students' confidence and feelings regarding success in the class.

Unit:	Date:	Lesson:	

Essential learning standard: List the essential learning standard for the unit addressed by today's lesson.

Learning objective: As a result of class today, students will be able to . . .

Essential standard for Mathematical Practice: As a result of class today, students will be able to demonstrate greater proficiency in which standard for Mathematical Practice?

Formative assessment process: How will students be expected to demonstrate mastery of the learning objective during in-class checks for understanding, teacher feedback, and student action on that feedback?

Probing Questions for Differentiation on Mathematical Tasks	
Assessing Questions	**Advancing Questions**
(Create questions to scaffold instruction for students who are "stuck" during the lesson or the lesson tasks.)	(Create questions to further learning for students who are ready to advance beyond the learning standard.)

Tasks	What Will the Teacher Be Doing?	What Will the Students Be Doing?
(Tasks can vary from lesson to lesson.)	(How will the teacher present and then monitor student response to the task?)	(How will students be actively engaged in each part of the lesson?)
Beginning-of-Class Routines How does the warm-up activity connect to students' prior knowledge, or how is it based on analysis of homework?		

Figure 2.24: Mathematical Practices lesson-planning tool.

continued →

Task 1 How will the students be engaged in understanding the learning objective?		
Task 2 How will the task develop student sense making and reasoning?		
Task 3 How will the task require student conjectures and communication?		
Closure How will student questions and reflections be elicited in the summary of the lesson? How will students' understanding of the learning objective be determined?		

Source: Kanold, Kanold, & Larson, 2012, pp. 53–54.

Figure 2.24: Mathematical Practices lesson-planning tool (continued).

Visit **go.solution-tree.com/mathematicsatwork** to download a reproducible version of this figure.

The How

Your team's work during the unit of instruction involves engaging one another in discussions about individual lessons. These lessons are focused on specific essential questions for the daily learning objective. If your team has used the before-the-unit-begins high-leverage team actions to plan the unit, then all elements needed to carry out the lesson are in place. Now the act of teaching and learning for the unit begins.

Understanding the Mathematical Practices Lesson-Planning Tool

At this point, there are some aspects of the Mathematical Practices lesson-planning tool that you and your collaborative team should contemplate to be successful. Consider the following frequently asked questions and responses regarding the Mathematical Practices lesson-planning tool.

Question: Is the Mathematical Practices lesson-planning tool appropriate for all mathematics content?

Answer: Yes. Each individual component of the tool is conducive to any mathematics content. In addition, during instruction for the unit, the structure this tool provides helps students come to know what to expect in a mathematics lesson and, hence, provides them some opportunity to note consistency and structure in *doing* the mathematics of the lesson.

Question: Will the Mathematical Practices lesson-planning tool work in a class of diverse learners?

Answer: Yes. When the Mathematical Practices lesson-planning tool asks, "What will the students be doing?", there is no focus on any particular characteristic of the student. In other words, the question addresses any student and every student—this includes consideration at the individual student level. So, the question is relative to your students and their characteristics. Other things, such as the structure of the classroom, also determine what students will be doing. For example, when students are working during the small-group instruction activities, what they are doing looks very different from students who are working individually or participating in discourse during whole-group instruction.

Question: We know the use of effective formative assessment strategies requires the collaborative team to make a commitment to plan an in-class common formative assessment process. Where do we see evidence of this process in the Mathematical Practices lesson-planning tool?

Answer: Every point of the lesson provides the opportunity for your assessment of student understanding and collection of information. For that data collection to become part of a formative assessment process for the student, every student must adjust and respond to his or her feedback and interaction with you or his or her peers during the lesson.

As such, every point of your lesson should provide a student opportunity to experience an effective formative assessment process. For example, the tool's "Assessing Questions" section provides an opportunity for teachers to help students scaffold their learning on a task or lesson to persevere and make progress during the lesson and the unit.

Question: Is the Mathematical Practices lesson-planning tool compatible with a specific mathematics curriculum or textbook series?

Answer: No. The Mathematical Practices lesson-planning tool supports all mathematics instructional designs, so it is compatible with any mathematics curriculum or textbook series.

Use the questions from figure 2.25 to work with your collaborative team to study the lesson-planning tool from figure 2.24.

Directions: Closely examine the Mathematical Practices lesson-planning tool (figure 2.24, pages 125–126). In your collaborative team, discuss how you can use the lesson-planning tool to help your current efforts to plan mathematics lessons for your grade level.

1. What are important factors to consider when planning lessons for an in-class formative assessment process?

2. How can you use the Mathematical Practices lesson-planning tool's assessing and advancing questions to facilitate small-group student-to-student discourse and differentiate learning around the higher-level-cognitive-demand mathematical tasks you use during the lesson?

3. How does the Mathematical Practices lesson-planning tool support the elements of your lesson—both the learning standards and the Mathematical Practices?

4. How will each mathematical task you choose for the lesson ensure students will be actively engaged in the mathematics?

5. Consider your start-of-lesson and end-of-lesson routines. How do these routines connect to the learning objectives of your lesson?

Figure 2.25: Learning about the CCSS Mathematical Practices lesson-planning tool.

Visit **go.solution-tree.com/mathematicsatwork** to download a reproducible version of this figure.

Lesson Study: Using the Mathematical Practices Lesson-Planning Tool for Collective Inquiry

During lesson study, your teacher team develops an expectation related to student learning for the lesson. You and your colleagues identify a challenging mathematical concept for students, and then design *together* a lesson to address the chosen essential learning standard. In the process of the lesson design, your team explores ideas about student learning as it relates to the chosen essential standard and decides what it is students are expected to *do* during the lesson.

A more informal model for promoting the idea of lesson study and a collective inquiry into your teaching practice is for one or two collaborative team members to enter lesson element ideas into the lesson-planning tool and present to the team for discussion and feedback. All team members subsequently contribute ideas and suggestions to revise the lesson.

When possible, observe the lesson you design as one team member teaches it, debrief about your observations, make changes to the lesson design, and then reteach the lesson with a final debriefing of the second instructional episode. It may seem time intensive for every unit, but try to commit to some type of collective inquiry or lesson study together at least once or twice per semester.

Our experience with middle school teacher teams is that lesson study will be one of the best work activities you can do together. The benefit of lesson study is the professional learning that results in your deep, collaborative discussions about content, mathematics instruction, and student learning. For more information and insight into how to do an effective lesson-study design together, visit **go.solution-tree .com/mathematicsatwork** to see additional lesson-design study models.

Remember, use a lesson-study lesson to develop your expectations for learning around increasing student proficiency in Mathematical Practices and processes. Choose a content standard you know to be problematic for students as part of your collective inquiry or lesson study, such as "Understand that positive and negative numbers are used together to describe quantities having opposite directions or values" (6.NS.1; NGA & CCSSO, 2010, p. 43). Then, use various resources to learn more about the content, connections to other mathematical concepts, and what research informs you about student learning of integers. By the end of the lesson study, you have gained in content knowledge and pedagogical knowledge and continually enhanced your professional knowledge as recommended in the professionalism principle of *Principles to Actions* (NCTM, 2014). You have also raised the level of respect and trust among team members. The lessons you learn and solution pathways you discover are valuable for your future lessons as well.

You can use the tool in figure 2.26 (page 130), or parts of the tool, to help your team with collaborative planning for a lesson study.

You can then use the observation tool in figure 2.27 (page 131) as a simple way to record data about what students are doing during the lesson that your team prepares and observes together. Remember that the intent of the lesson study, and collective inquiry as a team, is not so much to observe the teacher, as it is to observe the students and determine if they actually learned the standard via the collaborative tasks you chose for them to do that day. As Hattie (2012) states in the title of his book, *Visible Learning for Teachers*, the goal as you work together becomes more and more transparent.

Lesson-Design Components	Questions to Consider	Comments
Essential learning standard	What is the essential learning objective for the lesson? How does it connect to the essential learning standards for the unit?	
Assessment (formative, embedded, or summative)	What formative assessment strategies will we employ during the lesson? How will students self-assess their understanding and the understanding of their peers?	
Questioning	What are the assessing questions or prompts we can use to get students unstuck during the lesson? What are the advancing questions we can use to further student understanding? How will we engage students so that all students are required to think about the questions? How will we address a balance of cognitive-demand-level questions?	
Mathematical Practice	Which Mathematical Practice will students develop? How will they develop it? What evidence will we observe?	
Beginning-of-class routine	How do we address prior knowledge? How will all students be actively engaged in the opening activity?	
Activity or task one: How will students be engaged in understanding the essential learning standards?	What are all the ways the task can be solved? Which of these methods do you think students will use? What misconceptions might students show? What errors might students make?	
Activity or task two: How will the task develop student sense making and reasoning?	What is our expectation of student engagement for sense making and reasoning? How will we ensure the task is accessible to all students while still maintaining a high cognitive demand? What student-to-student interaction will we employ?	
Activity or task three: How will the task require student conjectures and communication?	How does the plan address orchestrating the class discussion? Which solution paths do we want shared during the class discussion? In what ways will the order in which solutions are presented help develop students' understanding of the mathematical ideas that are the focus of the lesson?	
Student-led closure	How do students know they met the learning target for the day? What evidence will we use to determine the level of student understanding for the essential learning standard?	

Figure 2.26: Mathematical Practices lesson-study planning and analysis tool.

Visit **go.solution-tree.com/mathematicsatwork** for a reproducible version of this figure.

Date: Course:	
Lesson Learning Standard: _____	
Lesson-Design Components	**Observations of Student Actions (Observed Data—What Are Students Doing?)**
Assessment (formative, embedded, and summative)	
Questioning	
Mathematical Practice and process	
Beginning-of-class routine	
Activity or task 1	
Activity or task 2	
Activity or task 3	
Student-led closure	

Figure 2.27: Lesson-study student data observation tool.

Visit **go.solution-tree.com/mathematicsatwork** to download a reproducible version of this figure.

For a more intensive lesson-design process, consider the completed three-day planning tool for a unit on exponential functions in figure 2.28 (page 132–133).

Unit: Rates and Ratios

Date: Days 6 through 8 of the unit

Lesson: Finding Unit Rates

Essential learning standard: 7.RP—Analyze proportional relationships and use them to solve real-world and mathematical problems.

1. Compute unit rates associated with ratios of fractions, including ratios of lengths, areas, and other quantities measured in like or different units. For example, if a person walks ½ mile in each ¼ hour, compute the unit rate as the complex fraction ½/¼ miles per hour, equivalently 2 miles per hour.

2. Recognize and represent proportional relationships between quantities.

    a. Decide whether two quantities are in a proportional relationship, for example, by testing for equivalent ratios in a table or graphing on a coordinate plane and observing whether the graph is a straight line through the origin.

    b. Identify the constant of proportionality (unit rate) in tables, graphs, equations, diagrams, and verbal descriptions of proportional relationships.

    c. Represent proportional relationships by equations. For example, if total cost t is proportional to the number n of items purchased at a constant price p, the relationship between the total cost and the number of items can be expressed as $t = pn$.

    d. Explain what a point (x, y) on the graph of a proportional relationship means in terms of the situation, with special attention to the points $(0, 0)$ and $(1, r)$ where r is the unit rate.

As a result of class today, students will be able to compute unit rates with ratios of fraction lengths and represent proportional relationships between quantities.

Formative assessment: How will students be expected to demonstrate mastery of the learning standard during in-class checks for understanding?

1. Communicate connections between ratios and graphing.

2. Recognize (orally, in writing, and/or by drawings) proportional relationships between quantities.

3. Describe (orally, in writing, and/or by drawings) differences between unit rates.

---

continued →

Probing Questions for Differentiation on Mathematical Tasks

Assessing Questions	Advancing Questions
(Scaffold instruction for students who are stuck during the lesson or the tasks.)	(Further learning for students who are ready to advance beyond the standard during class.)
1. What is a unit rate?	1. Is the time of 5 minutes and 20 seconds equal to 5.2 as a decimal? Why or why not?
2. Explain what 5:20 would be as a fraction—if a student uses 5.2 as a decimal.	2. How does this type of problem help you to understand why unit rates are helpful?
3. How do you calculate a unit rate?	3. Based on the data from the task that follows, create a graph to represent Mrs. Giuliano's time for swimming.
4. Why is a unit rate helpful?	
5. How does graphing help you identify if two quantities have a proportional relationship?	4. What did you need to consider in order to build the scale on the graph?
6. How does the unit rate help you write an equation to represent the proportional relationship?	5. How does graphing Mrs. Giuliano's data help in this situation and future situations?

Targeted Mathematical Practice

Mathematical Practice 1: Make sense of problems and persevere in solving them.

Mathematical Practice 3: Construct viable arguments and critique the reasoning of others.

(NGA & CCSSO, 2010, p. 6)

Activity or Task	What Will the Teacher Be Doing?	What Will Students Be Doing?
	The teacher will be observing, asking questions, responding to student questions, providing appropriate resources for students, and providing targeted support to students.	The students will be actively engaged in the lesson by collaborating in small groups, responding to teacher and peer questions and comments, asking questions, using the learning tools, and recording their work as instructed.
Beginning-of-Class Routines Prior to this lesson, students have explored ratios and unit rates. The expectation is that students are already familiar with calculating unit rates and representing a unit rate on a graph.	The teacher will pass out pictures of different images for students to calculate the unit rate. For example, it may be a photo of a sign that says "4 packs/$11.00 for Coca-Cola® products." The directions will be to calculate the unit rate. The purpose of this warm-up is to help connect to sixth-grade content while also reviewing vocabulary.	Students will work in their small groups to calculate the unit rates, discussing among themselves how to make the calculations. Students will also be encouraged to identify when there may be more than one unit rate that is helpful. For example, the ratio 4 packs/$11.00 could also be broken down to the cost per can if students know how many are in one pack.

Figure 2.28: Sample Mathematical Practices lesson-planning tool for grade 7.

continued →

Activity or Task 1 Students will engage in understanding the learning standard by providing information about their prior knowledge. Subsequently, the teacher will activate this prior knowledge as students discuss similarities and differences between the plane figures.	The teacher will distribute the task to groups of students who are seated heterogeneously. The teacher will instruct the students on how to work through the problems and the time frame, and remind students about group expectations. The teacher will introduce that the learning goal is for students to compute unit rates using fractions and apply unit rates to real-life scenarios. For groups or classes struggling to stay focused, the teacher will be prepared to scaffold the questions on the task.	Students will be discussing their thinking and asking each other questions as they work through each question in the task. If students are stuck or unsure of how to proceed, they will use their group for support first before seeking assistance from the teacher.
Activity or Task 2 This task will develop student sense-making and reasoning ability by requiring students to consider the responses of other students and to use this information to check their own understanding.	As the students work on the task, which addresses the skill of calculating unit rates while also providing a real-life context, the teacher will be circulating around the room prepared to ask questions and facilitate a meaningful conversation in each group.	Students will be listening to each other and asking for clarification where needed. It will be less about the answer and more about understanding the process and learning target.
Activity or Task 3 This task will require student conjectures and communication by promoting mathematical discourse that provides opportunities for debate and consensus.	As students wrap up their group work, the teacher will monitor group thinking and choose specific groups to share their thinking. As students are working, he or she may even ask certain students or groups to prepare their work on the board in preparation for the whole-group conversation.	Students will be accountable for communicating with each other in their groups, but they will also be accountable to attend to the group conversation and summary. This will allow groups to add or change thinking based on thinking from other groups.
Closure The teacher will elicit student questions and reflections in the summary of the lesson by using assessing and advancing questions. Students' understanding of the learning target will be determined by the teacher's assessment of students' interactions during group discourse and whole-class discourse.	The teacher will use the discussions she has observed to ask specific advancing questions of the class and assign homework for the night. The teacher will collect student self-reflections along with in-class tasks to look for concepts she will need to continue to build on during the next day.	Students will self-reflect on how they are doing in understanding how to calculate a unit rate in a real-life situation and why it is helpful to know how to do so.

continued →

7.RP.1 and 7.RP.2: Unit Rates and Proportionality Task

Mrs. Giuliano is training for an IRONMAN Race, which consists of swimming 3.8 km, biking 180.25 miles, and running 26.2 miles. She needs help estimating how long the swim will take her based on her previous race times. This past weekend she participated in a sprint triathlon. She completed a 400 m swim in 5 minutes and 20 seconds (5:20). The swim distance in an IRONMAN is 3.8 km.

1. How long would it take her to go ¼ of the distance of the sprint triathlon (400 m)? Describe how you figured this out.

2. If she were able to keep this constant speed for the IRONMAN, how long would it take her to swim 3.8 km (3,800 m)?

3. How did you calculate the answer to question 2? Explain your thinking.

4. After the sprint triathlon, Mrs. Giuliano competed in one more race before the IRONMAN. In this race, she swam 1,200 m in 16 minutes.

   a. Is this time proportional to her sprint time of 400 m in 5 minutes and 20 seconds? Show your work.

   b. What is another way you could prove they are or aren't proportional?

5. As you worked through all these problems, what are two strategies that helped you most?

Source for standards: NGA & CCSSO, 2010, pp. 6, 48.

Figure 2.28: Sample Mathematical Practices lesson-planning tool for grade 7 (continued).

Visit **go.solution-tree.com/mathematicsatwork** to download a reproducible version of this figure.

Based on your responses to the questions in figure 2.26 (page 130) and the sample middle school lesson in figure 2.28, there are important details to consider regarding the Mathematical Practices lesson-planning tool. First, take a close look at the assessing questions. These questions are particularly important for situations when students need intervention support or help getting unstuck so that they can develop a strong understanding of the mathematics. These questions can serve as a tool for scaffolding student or student team learning and for modeling ways students might develop their own questions or respond to your questions to promote action on their learning. These questions help your students create a culture of perseverance and give them confidence to engage in productive struggle as a positive outcome of class.

The advancing questions are necessary for students or student teams who might recall prior knowledge beyond the scope of the lesson or who otherwise are able to develop understanding more quickly than other students or student teams. The advancing questions raise the cognitive-demand level of the task but still keep students within the mathematical standards for the lesson.

Consider Mathematical Practice 3, "Construct viable arguments and critique the reasoning of others." In the lesson example in figure 2.28, opportunities for students to engage in Mathematical Practice 3 occur as students work through each activity and task and then share this information within their student team or within the whole-class discourse. The mathematical tasks you choose provide the foundation for small-group mathematical discourse that will most likely fuel your effort to promote students' agreements and disagreements in understanding about the plane figures and how they do or do not relate to each other.

Students' justifications of their understandings and their interpretations of the understandings of others are probable ways Mathematical Practice 3 comes into play during this lesson. You should work as a collaborative team to use either the Mathematical Practices lesson-planning tool *together*, or a more informal procedure at least once per unit—if not more often. As you become confident in the high-leverage team actions before and during the unit, you will find it easier to design, observe, reflect on, plan, and implement more lessons *together*.

Collaborative Planning and Collaborative Reflection

Collaborative planning is not the end to your team's work on lesson design. An essential final step is collaboratively reflecting on the success of the lesson as implemented in your collaborative team members' classrooms. Also important is what you can learn from each lesson and from each other to inform both upcoming lessons in future units as well as the design of a similar lesson for future implementation. Consider the reflection questions in figure 2.29.

Add your lesson reflections, including sample student solutions, to each lesson plan for the unit and keep them on file. Using an electronically shared folder will make these lesson notes readily available for future lesson planning. While this approach falls short of formal lesson study, such reflection provides ongoing opportunities for you and your team to continuously improve each lesson and learn from your experience.

Directions: With your collaborative team, answer the following prompts after implementing the lesson designed from the CCSS Mathematical Practices lesson-planning tool.

1. What level of student engagement with the lesson tasks did you observe? Describe how more direction, support, or scaffolding might have been provided if necessary.

2. Did students produce a variety of solutions or use a variety of strategies? If not, how might the structure of the tasks be redesigned to make them more open?

3. Describe any unexpected or novel solutions that would be worth remembering and incorporating into future classroom discourse.

4. What student misconceptions became evident during the lesson?

5. Which solutions were discussed and in what order? How did those choices support the classroom discourse? Upon reflection, might other choices have been more productive?

Figure 2.29: Reflection debriefing prompts to follow implementation of the CCSS Mathematical Practices lesson-planning tool.

Visit **go.solution-tree.com/mathematicsatwork** to download a reproducible version of this figure.

Creating a repository of collaborative lessons, including plans, mathematical tasks, sample student work, and reflections, is an essential team activity so each member continues to improve and refine his or her professional practice. Ultimately, this activity should increase the quality of your team's instruction and your students' learning. If you have not yet established such a repository, doing so is an important component of collaborative planning, no matter what type of lesson-planning tool your team uses.

Your Team's Progress

It is helpful to diagnose your team's current reality and action during the unit. Ask each team member to individually assess your team on the eighth high-leverage team action using the status check tool in table 2.4. Discuss your perception of your team's progress on using a lesson-design process for lesson planning and collective team inquiry. The real value in collaborating occurs in your discussions after you have tried the lesson. As your team seeks stage IV—sustaining—you increase the probability that all lessons contain an appropriate balance of higher- and lower-level-cognitive-demand mathematical tasks and provide opportunities for all students to benefit from your formative assessment lesson planning.

Table 2.4: During-the-Unit Status Check Tool for HLTA 8—Using a Lesson-Design Process for Lesson Planning and Collective Team Inquiry

Directions: Discuss your perception of your team's progress on the eighth high-leverage team action—using a lesson-design process for lesson planning and collective team inquiry. Defend your reasoning.			
Stage I: Pre-Initiating	**Stage II: Initiating**	**Stage III: Developing**	**Stage IV: Sustaining**
We do not use a lesson-planning tool.	We plan for instruction using the Mathematical Practices lesson-planning tool or other lesson templates independently.	We develop common lessons, either using the Mathematical Practices lesson-planning tool or other lesson templates but do not discuss the implementation.	We develop and implement common lessons at least once per unit, either using the Mathematical Practices lesson-planning tool or other lesson templates.
We do not know if our lessons provide for student demonstrations of understanding.	We discuss student demonstrations of understanding but do not have a common agreement on how to achieve them.	We collaboratively agree on how students should demonstrate understanding but we do not make instructional adjustments based on those agreements.	We ensure all lessons contain successful opportunities for students to demonstrate understanding.
We do not know about lesson study.	We have read about lesson study but do not create the time to do it.	We have engaged in a team lesson study but not as an ongoing practice.	We actively engage in a team lesson study once per unit and debrief in order to learn more about our students and to learn from each other.

Visit **go.solution-tree.com/mathematicsatwork** to download a reproducible version of this table.

Your team discussions about the daily lessons will ensure alignment to your unit learning standards and allow your team to make adjustments to in-class practices and student actions during the unit.

Setting Your During-the-Unit Priorities for Team Action

When your school functions as a PLC, your course-level collaborative team must make a commitment to pursue the three high-leverage team actions outlined in this chapter.

> HLTA 6. Using higher-level-cognitive-demand mathematical tasks effectively
>
> HLTA 7. Using in-class formative assessment processes effectively
>
> HLTA 8. Using a lesson-design process for lesson planning and collective team inquiry

As a team, reflect together on the stages you identified with for each of the three team actions. Based on the results, what should be your team's priority? Use figure 2.30 (page 140) to focus your time and energy on actions you deemed most urgent in your team's preparation and reflection during the unit. Remember, always lesson plan from the students' point of view. The following questions are at the heart of your during-the-unit work with your team: What will your students be doing during every aspect of the lesson? How will you provide formative feedback to them? How will they be expected to take action on that feedback? Focus on these few, but complex tasks, and work to integrate them into the planning and delivery of lessons for you and your team.

Remember the three high-leverage team actions in this chapter reflect steps two and three of the PLC teaching-assessing-learning cycle (see figure 2.1, page 80) and will help you prepare for the challenge of teaching and learning during the unit. They are also linked to teacher actions that will significantly impact student learning in your class.

Our attention now turns to chapter 3, and to steps four and five of the PLC teaching-assessing-learning cycle, with a focus on implementing a formative assessment process in response to the end-of-unit common assessments, your team's and your students' formative response to the data revealed at the end of the unit, and the impact on instruction during the next unit of study.

Directions: Identify (circle) the stage you rated your team for each of the three high-leverage team actions, and provide a brief rationale.

6. Using higher-level-cognitive-demand mathematical tasks effectively

Stage I: Pre-Initiating   Stage II: Initiating   Stage III: Developing   Stage IV: Sustaining

Reason: _____

_____

_____

_____

_____

7. Using in-class formative assessment processes effectively

Stage I: Pre-Initiating   Stage II: Initiating   Stage III: Developing   Stage IV: Sustaining

Reason: _____

_____

_____

_____

_____

8. Using a lesson-design process for lesson planning and collective team inquiry

Stage I: Pre-Initiating   Stage II: Initiating   Stage III: Developing   Stage IV: Sustaining

Reason: _____

_____

_____

_____

_____

With your collaborative team, respond to the red light, yellow light, and green light prompts for the high-leverage team actions that you and your team believe are most urgent.

Red light: Indicate one activity you will stop doing that limits effective implementation of each high-leverage team action.

Yellow light: Indicate one activity you will continue to do to be effective for each high-leverage team action.

Green light: Indicate one activity you will begin to do immediately to become more effective with each high-leverage team action.

Figure 2.30: Setting your collaborative team's during-the-unit priorities.

Visit **go.solution-tree.com/mathematicsatwork** to download a reproducible version of this figure.

CHAPTER 3

After the Unit

You can't learn without feedback. . . . It's not teaching that causes learning. It's the attempts by the learner to perform that cause learning, dependent upon the quality of the feedback and opportunities to use it. A single test of anything is, therefore, an incomplete assessment. We need to know whether the student can use the feedback from the results.

—Grant Wiggins

You have just taught the unit and given your common end-of-unit assessment. Did students reach the proficiency targets for the unit's essential learning standards? How do you know? How do your students know? More importantly, what are the responsibilities of your collaborative team after the unit ends?

Your after-the-unit high-leverage team actions support steps four and five of the PLC teaching-assessing-learning cycle (see figure 3.1).

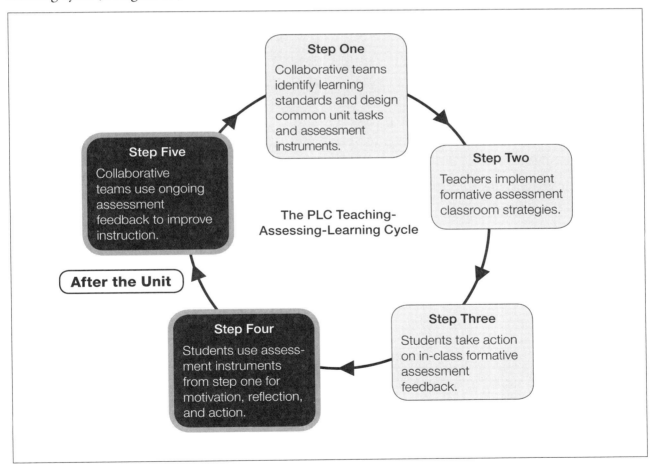

Source: Kanold, Kanold, & Larson, 2012.

Figure 3.1: Steps four and five of the PLC teaching-assessing-learning cycle.

Think about the last time you passed back an end-of-unit assessment. Did assigning the student a score or grade motivate the student to continue to learn and use the results as part of a formative assessment process? Did the process of learning the essential standards from the previous unit stop for the student as the next unit began? In a PLC culture, the answers are simple: the process of student growth and demonstrations of learning *never* stops.

Recall that in step one (chapter 1), your collaborative team unpacked and paced the essential learning standards for the unit, identified meaningful mathematical tasks to balance cognitive demand, created the common assessment instruments for the unit with scoring rubrics, agreed to proficiency targets for the essential learning standards, and agreed to common homework assignments that support students' conceptual understanding and procedural fluency.

In steps two and three (chapter 2), the focus was on what students were doing in class—how your collaborative team developed and implemented lessons that encouraged students to demonstrate the Standards for Mathematical Practice and incorporated the effective use of formative assessment processes. You used the lesson-planning tool to organize the mathematical tasks you developed with chapter 1 into class-ready lessons.

Now, in step four of the PLC teaching-assessing-learning cycle, you investigate the process of consistent and accurate scoring of the common assessment instrument for immediate and constructive feedback to guide student learning. You examine the role your students play in using your feedback to set new proficiency targets and take action on their learning progress.

Finally, in step five, you will consider ways in which you work as part of your collaborative team—reflecting on student performance during the unit, adjusting your decisions about the next unit based on that information, and taking action as needed based on the end-of-unit assessment results in order to inform "feedback to students, instructional decisions, and program improvement" as outlined in NCTM's assessment principle in *Principles to Actions* (NCTM, 2014).

To complete the analysis expected in steps four and five, you and your collaborative team engage in the final two high-leverage team actions.

> HLTA 9.   Ensuring evidence-based student goal setting and action for the next unit of study
>
> HLTA 10. Ensuring evidence-based adult goal setting and action for the next unit of study

As Hattie (2012) states:

> My role as a teacher is to evaluate the effect I have on my students. . . . This requires that teachers gather defensible and dependable evidence from many sources, and hold collaborative discussions with colleagues and students about this evidence, thus making the effect of their teaching visible to themselves and to others. (p. 19)

The PLC teaching-assessing-learning cycle allows teacher and student reflection around evidence of learning on a unit-by-unit basis throughout the year and provides for a progression of learning beyond the end of the unit.

HLTA 9: Ensuring Evidence-Based Student Goal Setting and Action for the Next Unit of Study

Indeed, the whole purpose of feedback should be to increase the extent to which students become owners of their own learning.

—Dylan Wiliam

Reflect on your last end-of-unit assessment. What was the typical student response to the feedback you provided? How did students use this feedback to further their understanding of the content? Were they lined up at your door the next day for tutoring? Or was their response, "I got a _____. Better luck next time." Wiliam (2011) states it like this: "As soon as students get a grade on a test, the learning stops" (p. 123). Learning should not stop when your collaborative team provides meaningful assessment feedback to students; their score is actually the least-important aspect of the assessment. Students should reflect on their assessment performance and take action on their results. Using the assessment results, collaborative teams can re-engage students to continue the learning.

Wiliam (2007) makes the distinction between using assessment instruments for the purposes of (1) monitoring, (2) diagnosing, or (3) formatively assessing. He states:

> An assessment monitors learning to the extent it provides information about whether the student, class or school is learning or not; it is diagnostic to the extent it provides information about what is going wrong, and it is formative to the extent it provides information about what to do about it. (p. 1062)

Thus, a key formative assessment feature after the unit ends will be the process you have in place for your students to use and respond to the results of their assessment performance.

Once again, recall there are four critical questions every collaborative team in a PLC asks and answers on an ongoing, unit-by-unit basis.

1. What do we want all students to know and be able to do? (The essential learning standards)

2. How will we know if they know it? (The assessment instruments and tasks teams use)

3. How will we respond if they don't know it? (Formative assessment processes for intervention)

4. How will we respond if they do know it? (Formative assessment processes for extension and enrichment)

High-leverage team action 9—ensuring evidence-based student goal setting and action for the next unit of study—guarantees your students use common end-of-unit assessment results as part of a formative assessment process that responds to critical questions three and four in support of NCTM's (2014) assessment principle. What is the expected after-the-unit response if students demonstrate there are certain essential learning standards they do or do not know? Answering this question, and making appropriate instructional adjustments and putting necessary student supports in place, are key characteristics of professional teachers of mathematics as outlined in NCTM's (2014) professionalism principle.

High-Leverage Team Action	1. What do we want all students to know and be able to do?	2. How will we know if they know it?	3. How will we respond if they don't know it?	4. How will we respond if they do know it?
After-the-Unit Action				
HLTA 9. Ensuring evidence-based student goal setting and action for the next unit of study			▢	▢

▢ = Fully addressed with high-leverage team action

The What

In many classrooms, students typically receive their assessment back from their teacher, look at their score, become excited, sad, or mad (depending on the score), and place the assessment in their notebook or backpack never to be seen again (unless it was a good score, which means it may end up on the refrigerator at home). This action of locking away the assessment wastes a valuable formative learning opportunity for your students.

Your team's work on this high-leverage team action will diminish that response. Your middle school students will reflect on successes and focus their next steps based on evidence of weakness if the assessment instrument encourages formative student learning.

All students in your class should see assessment results as a means to better understand their current mathematical knowledge and be able to use the information on the common end-of-unit assessment to improve their mathematical understanding. Thus, assessment is an integral part of the learning process for students, which is why it is in the middle of the PLC teaching-assessing-learning cycle and used from unit to unit.

The value of using assessment results includes two very important elements. First, students need to be given FAST feedback on their work in order to improve their mathematical understanding. Second, students must have the tools and the opportunity to learn how to use their end-of-unit assessment results in formative ways to take action on your feedback.

The How

As a first step, brainstorm ways you and your collaborative team might involve students in a formative assessment process that includes feedback and action on the previous units' results. Make a list of your ideas, and return to them as you continue through this chapter.

To be part of a formative learning process for students, your feedback needs to require student action. However, it also needs to be meaningful and effective. Recall our discussion during HLTA 7 (page 99) regarding the criteria for effective feedback.

Characterizing Effective Feedback: FAST—Fair, Accurate, Specific, and Timely

In your in-class formative assessment work with high-leverage team action 7, you learned the essential characteristics of effective in-class feedback (Reeves, 2011). These same components of effective feedback hold true for your work with the end-of-unit assessments.

1. **Feedback should be fair:** Effective feedback on the test rests solely on the quality of the student's demonstrated work and not on other characteristics of the student.

2. **Feedback should be accurate:** Effective feedback on the test acknowledges what students are currently doing well and correctly identifies errors they are making. According to Stephen Chappuis and Rick Stiggins (2002), "Effective feedback describes why an answer is right or wrong in specific terms that students understand" (p. 42).

3. **Feedback should be specific:** Your test notes and feedback "should be about the particular qualities of [the student's] work, with advice on what he or she can do to improve, and should avoid comparison with other pupils" (Black & Wiliam, 2001, p. 6). Try to find the right balance between being specific enough that the student can quickly identify the error or logic in his or her reasoning but not so specific that you do the correction work for him or her. Does your test feedback help *students* correct their thinking as needed?

4. **Feedback should be timely:** Effective feedback on the end-of-unit assessment must be provided in time for students to take formative learning action on the results before too much of the next unit has taken place. As a general rule, you should pass back end-of-unit assessments and the results for proficiency to students within forty-eight hours of the assessment.

To begin your team's development of using feedback effectively, practice using a sample assessment task from chapter 1—figure 1.21 (page 56)—the sand task. The task is presented again in this chapter as figure 3.2 (page 146).

Sand Task
Jamal is filling bags with sand. All of the bags are the same size. Each bag must weigh less than 50 pounds. One sand bag weighs 57 pounds, and another sand bag weighs 41 pounds. Explain whether Jamal can put sand from one bag into the other so that the weight of each bag is less than 50 pounds.

Example of Good Work—Student One:

Jamal can pour Sand from one bag into the other. He has to pour 8 pounds from the sand bag that weighs 57 lbs to the sand bag that weighs 41 lbs That way each sand bag weighs less than 50 lbs.

$$57 - 8 = 49$$
$$41 + 8 = 49$$
$$50 > 49$$

Example of Not-as-Good Work—Student Two:

Jamal can pour 7 pounds of sand from the 57 pound bag into the 41 pound bag. The 2 bags will weigh 50 pounds and 48 pounds.

Jamal can

Source: Task reprinted from Smarter Balanced Assessment Consortium, n.d. Used with permission.

Figure 3.2: Sample grade 6 task.

Visit **go.solution-tree.com/mathematicsatwork** to download a reproducible version of this figure.

As an exercise in providing accurate and specific student feedback, you and your collaborative team should discuss the type of feedback you would give to the students whose responses are shown in figure 3.2.

Each team member should record his or her feedback to the students using the discussion questions from figure 3.3 and then discuss.

Directions: Record your feedback to the students whose responses appear in figure 3.2. Then discuss with your team the type of feedback you would provide to the student.

1. How did the strategies the student used demonstrate his or her understanding of the essential learning standard the task assesses?

2. How did the feedback you generated build on the student's strengths to address the learning needs for the task?

3. How did the feedback you generated guide the student to understand an error (if you believe there was an error) without being too directive?

4. How did the feedback you generated recognize student effort and accuracy?

5. How did the feedback recognize the Mathematical Practices the student used for the task?

6. In what ways did you provide the feedback in enough time to allow for effective student action?

Figure 3.3: Collaborative team student feedback discussion question tool.

Visit **go.solution-tree.com/mathematicsatwork** to download a reproducible version of this figure.

Increasing Accuracy for and Decreasing Variance of Feedback

You will have variability in how your team members score similar items unless you come to an agreement on the scoring and accuracy of the feedback you give to students. If you do not come to an agreement, grading inequity will occur within the course.

You can use the reflection activity in figure 3.4 (page 148) to gather information on your various perspectives for each test question or task. Although this is best done as a before-the-unit team activity, it's very possible that you might not have had time to do so, in which case, this activity will help produce equity in feedback.

Directions: Answer the following questions with your collaborative team.	
Scoring Rubric Prompts	**Reflection**
1. What percentage of the exam assesses procedural fluency? What score would you assign to the procedural fluency questions? Are these questions worth more than one point? Would you give partial credit? When and why?	
2. What percentage of the exam assesses conceptual understanding? What score would you assign to the conceptual understanding questions? Would you give partial credit? When and why?	
3. What is the total possible score of the assessment? Did you each use the same total point score?	
4. If you assign more than one point to an item, does the scoring rubric clarify why students would receive specific points for a given student response or misconception?	
5. Do the evidence of student learning and the points for each essential learning standard reflect the instructional emphasis of the unit?	

Figure 3.4: Team reflection activity for group scoring an assessment.

Visit **go.solution-tree.com/mathematicsatwork** to download a reproducible version of this figure.

Figure 3.5 shows a portion of a scoring rubric created for an eighth-grade geometry unit assessment. As you review the example, do you agree with the team recommendations? If you were grading student work for this assessment, would these notes be sufficient information to score accurately and equitably from teacher to teacher on your team?

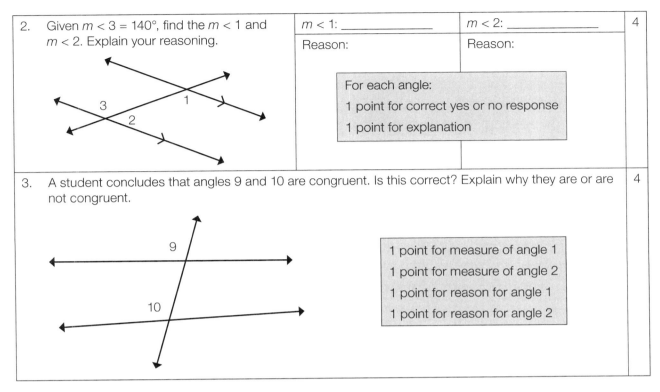

2. Given *m* < 3 = 140°, find the *m* < 1 and *m* < 2. Explain your reasoning.	*m* < 1: _____ Reason: For each angle: 1 point for correct yes or no response 1 point for explanation	*m* < 2: _____ Reason:	4
3. A student concludes that angles 9 and 10 are congruent. Is this correct? Explain why they are or are not congruent.	1 point for measure of angle 1 1 point for measure of angle 2 1 point for reason for angle 1 1 point for reason for angle 2		4

Source: Adapted with permission from Schaumburg SD 54, Schaumburg, Illinois.

Figure 3.5: Unit test scoring example.

Once your team has agreed on the scoring accuracy of the common assessments, you then set up a process to engage students in the assessment cycle. A critical action for students, which you should require, is to use the feedback from the common assessment to improve their learning. Students use your feedback (they pay attention to it and take action) to deepen their own mathematical understanding.

Select your most recent common end-of-unit assessment. Pull two to three examples of student work from the assessment, and share the feedback from each team member. Use the questions from figure 3.4 to consider how accurate your feedback is across each team member. What can your team do to continue progress in the specific, accurate, and fair feedback you offer to students?

Be sure to also discuss the timeliness of your feedback to students. The use of accurate and specific feedback will not help students unless it is also *timely*. For example, if a student only provides one way to solve a problem, you can provide feedback hints on the assessment task that could guide the student to consider another approach, such as, "What related facts do you know that could help you?" or "How could you use a graphing strategy?" However, if you provide this feedback to the student well after you give the assessment, your feedback will have minimal impact.

As a team, discuss your current practice for providing immediate feedback to your students on the end-of-unit assessment. Is it timely and effective so students can take action?

Student Action on Common End-of-Unit Assessment Feedback

When you return student work with your feedback on the end-of-unit assessment, your students need to be able to self-assess if they are meeting the proficiency targets set for each of the essential learning standards for the unit. Your feedback will allow students, with your support, to engage in activities that lead to progress on the essential learning standards for that unit as well as for the course.

How can this be done in your school or on your team for each unit-by-unit cycle? Consider the following two actions.

1. Create a student goal-setting reflection process to identify errors, and use the assessment results to form a plan.

2. Create a process for students to act on their plan and take action (allow them to improve their score for each essential learning standard on the end-of-unit assessment).

Simply assigning a grade to an assessment is the least-useful strategy to motivate students to further their understanding (Kanold, Briars, Asturias, Foster, & Gale, 2012). Students must use the feedback from the unit assessment as a formative learning opportunity, instead of a "test once, then done" approach. Your feedback will allow students to reflect on their strengths and challenges from the assessment. When you pass the test back in class, give your students time to self-assess their results, and make a plan of action to retool their learning. Wiliam (2011) describes that effective feedback should:

> Cause thinking: It should be focused; it should be related to the learning goals that have been shared with students; and it should be more work for the recipients than the donor. Indeed, the whole purpose of feedback should be to increase the extent to which students are owners of their learning. (p. 132)

To ensure student ownership of their learning, your team's feedback to students should be more than a simple grade. Figure 3.6 is a sample student self-assessment for the seventh-grade unit assessment on expressions outlined in figure 1.20 on page 47. Students complete the self-assessment page during class—preferably the day after the assessment. As you score each student's assessment, you indicate the points he or she earned on each item and provide feedback as needed. Then, when students receive their assessment score in class, each student can calculate the points he or she earned and the percent correct for each learning target and essential standard for the unit. Once students have completed their reflection, they then must complete specific and targeted actions to ensure they close the gap on each essential learning standard for which they are not proficient. Note that this teacher team thought the test should be worth 53 points. Determine if your team agrees.

Grade 7: End of Unit 2—Expressions Student Self-Assessment and Reflection

Directions: For student error analysis, use the following code system—Concept Error (CE), Silly Mistake (SM), Calculation Error (Calc).

7.EE.1: I can apply properties of operations as strategies to factor and expand linear expressions with rational coefficients to generate equivalent expressions.

7.EE.2: I can rewrite expressions in different forms to show how quantities are related.

Test Questions	Responses for Full Credit	Score	Cluster Score	Cluster Percent	Student Error Analysis
1	Distribute 11 into s and 9 correctly.	/1	/16		
2	Distribute $-\frac{2}{5}$ into x and 25 correctly.	/1			
3	Distribute correctly. Combine like terms. Give correct explanation.	/3			
4	Simplify expressions correctly with work. Give correct explanation.	/3			
5	Identify the correct mistake. Fix the error. Provide description of what the student did wrong.	/3			
6	Explain agreement or disagreement and why.	/2			
7	Show work. Choose correct answer. Explain why.	/3			
Student Written Reflection					

Figure 3.6: Sample student self-assessment—end-of-unit test cover page.

continued →

7.EE: Solve real-life and mathematical problems using numerical and algebraic expressions and equations.

7.EE.3: I can solve real-life and mathematical problems using operations with rational numbers in any form.

7.EE.4: I can use variables to represent quantities in a real-world or mathematical problem, and construct simple equations to solve problems by reasoning about the quantities.

Test Questions	Responses for Full Credit	Score	Cluster Score	Cluster Percent	Student Error Analysis
8	Ryan's expression. Alex's expression. A. Correct answer. Explanation using mathematical vocabulary B. Correct answer. Explanation using mathematical vocab. C. Answer with a detailed explanation.	/8	/37		
9	A. Correct expression in simplified form. B. Correct set-up of the equation. Correct value with a label.	/3			
10	A. Correct expression including the missing sides. Expression in simplified form. B. Correct explanation that includes adding all sides together, including needing to calculate the missing sides and combining like terms. C. Correct substitution of $a = 2$ and	/6			
11	A. Correct expression. Correct value substituted for v. B. Correct answer. Detailed explanation.	/4			
12	A. Correct answer. Valid explanation. B. Correct expression. C. Correct answer with work. D. Correct answer. Valid explanation for answer. E. Correct answer. Valid explanation.	/8			
13	Subtract 5 on both sides. $x = -12$	/2			
14	Multiply by 3 on both sides. $x = 18$	/2			
15	Distribute 3 into $(x - 2)$. Add 6 to both sides. Divide both sides by 3. $x = -2$	/4			
Student Written Reflection					
Unit Test Total		/53	%		

Source for standards: NGA & CCSSO, 2010, p. 49.

To facilitate this action-step process, your collaborative teams will need to create re-engagement strategies for students that align with the essential learning standards. What would re-engagement look like after the unit of instruction? Re-engagement is different and more powerful than reteaching.

Reteaching implies you teach content the same way as the first learning experience; when you reteach concepts, you often do not use student misconceptions from the unit assessments to engage thinking about the concepts in a new way. *Re-engagement* is a system of targeted intervention strategies that utilize data regarding student misconceptions to engage students differently in mathematical concepts. Effective collaborative teams review student misconceptions and craft re-engagement lessons and tools that directly tie to the learning standards on the assessment (Foster, 2008).

Thus, your team sets up processes that allow students to:

1. Reflect on the feedback from the unit assessment (usually when you pass back tests in class)

2. Use the specific feedback for reflection, motivation, and action (re-engagement)

3. Receive an improved score based on the assessment results (grades reflect actual learning of the standards)

Figure 3.7 (page 154) is a specific sample of a sixth-grade student self-assessment action plan for use during the unit and completed after students review your end-of-unit assessment feedback. Notice that students reflect on their progress throughout the unit with *checkpoint quizzes*—small formative assessments within the unit—that provide students with feedback to ensure they are monitoring their level of understanding *before* the common end-of-unit assessments.

Student Self-Assessment

Name: _____

Sixth Grade: Expressions Unit   Period _____   Date _____

For each essential learning standard, record how many questions of each learning standard you earned full credit on, and then decide how well you understand the learning standards at this time. For the final assessment, record how many points you earned, and determine what kinds of mistakes you made and the target level of mastery for each learning standard.

| Checkpoint quizzes are given to monitor learning and to address misconceptions with tutoring before the unit test. | | Point Quiz | | | Unit Test | | | | | | | |
| | | Expressions Readiness Check — How well do I know it? | | | Level of Accuracy — Why did I not earn full credit? | | | Level of Mastery | | | | |
Standards	Learning Target I can . . .	Points Earned	Percentage	Tutoring Yes or No? (Circle one.)	Questions on Test	Points Earned	Percentage	Exceed 90–100%	Proficient 70–89%	Approaching 50–69%	Some Evidence 25–49%	No Evidence 0–24%
6.EE.1	Evaluate numerical expressions involving whole number exponents.	6		Yes 0–69% No 70–100%	1–3	6						
6.EE.2	Translate words to mathematical expressions.	6		Yes 0–69% No 70–100%	4–6	6						
6.EE.3	Apply the properties of operations to create equivalent expressions.	6		Yes 0–69% No 70–100%	7–9	6						

For each learning target you are approaching, create an action plan of how you will learn the standards, and be prepared to take the recovery quiz (your second attempt at demonstrating mastery). Please remember to also write down times and dates. The more specific you are about your plan, the more likely you are to stick to it. "Whenever" is not specific!

| Action Plan — Concept mastery options I will attempt prior to taking a recovery quiz (Check all that apply, and write dates and times: morning, lunch time, advisory, or after school.) | | | | | | | | | |
Standards	Learning Target I can . . .	Standard Recovery Tutoring With Teacher	Standard Recovery Tutoring in MASC	Standard Recovery Tutoring With Another Teacher or Student	Standard Recovery by Independently Completing Review	Standard Recovery by Rereading Notes	Standard Recovery by Online Program	Date of Recovery Quiz	Score
6.EE.1	Evaluate numerical expressions involving whole number exponents.								
6.EE.2	Translate words to mathematical expressions.								
6.EE.3	Apply the properties of operations to create equivalent expressions.								

Student signature: _____   Date: _____

Parent signature: _____   Date: _____

Source for standards: NGA & CCSSO, 2010, pp. 60, 64.

Figure 3.7: Sample student self-assessment, goal-setting, and action plan.

As students reflect on their current reality throughout the unit, they create a plan to take action on standards they do not understand *prior* to the end-of-unit assessment. Students do not have a choice about whether or not to re-engage; learning is not optional. Your team ensures that these Tier 2–type RTI re-engagement structures for students are required.

The intent of students self-assessing their performance on the end-of-unit assessment is to help each student build responsibility for his or her own learning. While each student takes ownership for his or her individual progress on the learning standard, students may still work together to meet those standards. Student-to-student feedback is a vital component of the entire educative process, and the goal setting at the end of each unit is part of that student collaboration. This collaboration with peers and ownership of the learning pathway will engage students more deeply in the learning process and provide evidence for each student that effective effort is the route to mathematical understanding and success. Visit **go.solution-tree.com/mathematicsatwork** for additional sample action plans.

As you work with students to help them identify strengths, weaknesses, and the essential learning standards they still need to meet, what structures will you use to measure their progress? How will students work during the next unit to meet the learning standards they still do not know? They will need your help to take small steps toward reaching the learning standards from the previous unit. You may need to make a chart like the one in figure 3.8 (page 156) for each student to record his or her current proficiency on each essential learning standard for future progress. You might choose designations, like levels, for students to achieve that have predetermined descriptions of understandings to be reached. Alternatively, ask students to keep a journal or personal chart to record their own progress during the year.

For students with misconceptions identified through data on the end-of-unit assessment, your team could also develop a plan based on what works best for upcoming instruction. For example, if the next unit builds on the instruction from the current unit, tasks in the next unit should include small-group learning opportunities in which students can continue to work to address their misconceptions, developing proper prerequisite knowledge for the next unit. However, if the next unit contains new mathematical content standards, then the team needs to consider other options to support student learning.

This collaboration with peers and ownership of the learning pathway will engage students more deeply in the learning process and provide evidence for each student that effective effort is the route to mathematical understanding and success—an important lesson to learn early in life.

Valerie Tomkiel at Stevenson High School District 125 discussed the work of her algebra team (she is the team leader who helps to lead the middle school teachers in her district whose students feed into her high school). We share that discussion here on the feature box on page 157 in order to give you a glimpse into the nature and focus of her team's work and effort (V. Tomkiel, personal communication, May 2014).

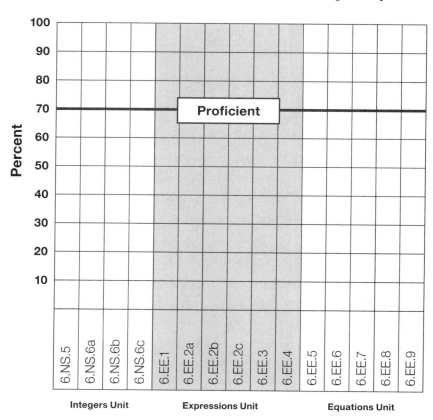

Figure 3.8: Student progress-tracking chart.

What makes for a good homework assignment?

- A good homework assignment has practice of the current learning target as well as some review of old material or prerequisite skills practice. The old material on homework helps the students to retain the information. Prerequisite skills can be helpful to get students prepared for an upcoming lesson. For example, before learning rational exponents, it may be helpful to practice adding, subtracting, and multiplying fractions.

- Good homework should include some written explanation or interpretation. This can be helpful to the teacher to assess where the students are in the learning process.

- Good homework includes the answers so the students can check their work as soon as they finish a problem.

How does the team reflect on the test results after the test?

- After a unit test, we share our mean percentages with each other. We also share the most common problem our students have gotten incorrect. We use this information to determine how prepared the students are for subsequent chapters. For example, if the student exams showed that students had not mastered writing equations of lines, as a team, we chose to continue to reteach this topic during the next chapter.

How are common assessments useful for the team?

- They require teachers to align their lessons to the pre-established learning targets.

- They allow us to compare our results—apples to apples. Our conversations about results are specific and helpful to teachers.

- Students benefit from a better quality test because there are more eyes to look it over.

How do you know an assessment is worthwhile? What makes it good?

- It measures students' progress on a specific standard.

- It is divided up by learning target so the student can use the information from the assessment to understand what he or she still needs to learn.

- It is timely. Students take the assessment after they have had time to process the information, and they receive feedback in a timely fashion so they can improve their understanding if necessary.

What is the team doing to foster a continuous improvement mindset for students?

- The team gives students remediation packets after quizzes. The students can retake quizzes prior to the unit test.

What does remediation look like?

- In our course, it means a lot of things depending on the situation. Each teacher has a plan for a student to remediate. Some things that are commonly used for remediation are redoing homework, completing additional practice worksheets, meeting with the teacher or ILC tutor, making quiz corrections, and watching and modeling problems from an online video.

Your Team's Progress

It is helpful to diagnose your team's reality and action after launching the unit. Ask each team member to individually assess your team on the ninth high-leverage team action using the status check tool in table 3.1. Discuss your perception of your team's progress on ensuring evidence-based student goal setting and action for the next unit of study. It matters less which stage your team is at and more that you and your team members are committed to working together and understanding how to support students who still need help after the end-of-unit assessment as your team seeks stage IV—sustaining.

Your careful, honest, and open answers on the status check tool will allow for movement toward effective implementation of formative assessment strategies to assist students and teachers alike in the learning process. After reflecting on some of the work you and your team have done in chapters 1 and 2, you must now turn to the final high-leverage team action, HLTA 10: Ensuring evidence-based adult goal setting and action for the next unit of study.

Table 3.1: After-the-Unit-Ends Status Check Tool for HLTA 9—Ensuring Evidence-Based Student Goal Setting and Action for the Next Unit of Study

Directions: Discuss your perception of your team's progress on the ninth high-leverage team action—ensuring evidence-based student goal setting and action for the next unit of study. Defend your reasoning.			
Stage I: Pre-Initiating	**Stage II: Initiating**	**Stage III: Developing**	**Stage IV: Sustaining**
We do not discuss whether our test feedback is fair, accurate, specific, or timely.	We discuss how our test feedback should be fair, accurate, specific, and timely, but we do not know what other team members actually do.	We provide fair, accurate, specific, and timely feedback to students but do not discuss its impact as a collaborative team.	We provide fair, accurate, specific, and timely feedback to students, and we discuss the impact of this feedback as a collaborative team.
We do not provide students with opportunities to respond to the feedback from the end-of-unit assessment.	We provide some constructive feedback to students on the end-of-unit assessment, but we do not require them to respond to the feedback.	We require students to correct their errors on the end-of-unit assessment.	We require students to correct their errors and identify the learning standards that are strengths and weaknesses.
We do not know the type of end-of-unit assessment feedback other team members use.	We do not have a team process in place for student response to the end-of-unit assessment results.	We work with each student to identify a plan for improvement and action based on end-of-unit results and improvement.	We work with each student to complete and carry out a plan for improvement and action based on end-of-unit results.

Visit **go.solution-tree.com/mathematicsatwork** to download a reproducible version of this table.

HLTA 10: Ensuring Evidence-Based Adult Goal Setting and Action for the Next Unit of Study

[High-impact teaching] requires that teachers gather defensible and dependable evidence from many sources, and hold collaborative discussions with colleagues and students about this evidence, thus making the effect of their teaching visible to themselves and to others.

—John Hattie

The final high-leverage team action and second after-the-unit pursuit is to consider how your team uses the results of the end-of-unit assessment. While it is important for students to use feedback to monitor and improve their understanding of mathematics, it is just as important for your team to reflect on the unit assessment data.

The What

You and your team need to use the unit's assessment instrument results to monitor and evaluate your instruction—actions that are at the heart of both the assessment and professionalism principles of *Principles to Actions* (NCTM, 2014). This provides a formative assessment learning process for your team, not just for the students. This tenth high-leverage action ensures your team reaches clarity on how to effectively respond *after* the unit ends, addressing the third and fourth PLC critical questions: How will we respond if they don't know it? How will we respond if they do know it?

High-Leverage Team Action	1. What do we want all students to know and be able to do?	2. How will we know if they know it?	3. How will we respond if they don't know it?	4. How will we respond if they do know it?
After-the-Unit Action				
HLTA 10. Ensuring evidence-based adult goal setting and action for the next unit of study			☐	☐

 ☐ = Fully addressed with high-leverage team action

Using the evidence from the common end-of-unit assessment, consider the following questions on your own and with your team:

- What went well in the unit?

- How well did students understand the essential learning standards of the unit?

- Which students need additional time and support to become proficient?

- Which students would benefit from an extension of the standard due to demonstrated proficiency?

- How well did we provide feedback to the students during the unit?

- How did the results vary by teacher? What are areas that warrant attention and need improvement for individual teachers as well as the team?

- How will we respond to the evidence of learning standards that did not result in student success?

The answers to these questions should be based on personal reflection and supported by the data from the common end-of-unit assessment. At this moment—step five in the PLC teaching-assessing-learning cycle (figure 3.1, page 141)—your team is charged with timely reflection on how your efforts during the unit did or did not meet with success. It is important to reflect before your team reaches too deeply into the next unit of instruction if the results are to have an impact on the instruction of the next unit.

Your collaborative team's unit assessment has its greatest payoff in this final high-leverage team action—using the student-performance results to make future instructional decisions for the next unit together. This supports using assessment results to inform "instructional decisions and program improvement" (NCTM, 2014, p. 89). From a practical point of view, this sets up about ten cycles of planning and reflection for the year. This may seem too often at first, but with practice it becomes the norm in a PLC at Work culture.

Your team will need to focus on the messages contained only within the data and not allow other factors to influence your reflection. This is a key feature of the effective use of the common unit assessment instrument—how your team data reflect on student acquisition of the essential learning standards for the unit.

The How

Return to the most recent common end-of-unit assessment you and your team implemented. Consider the results of the end-of-unit assessment and what questions the data surfaced regarding the essential learning standards for the unit. Respond to the questions in figure 3.9 based on an objective review of the results. Your team will need to focus on the messages contained *only* within the data and not allow other factors to influence your reflection. This is a key feature to the effective use of the common assessment: how the data reflect student acquisition of the essential learning standards for the unit. As you respond to the questions, make sure your review is objective and bias free. Often, when looking at data, we can make inferences beyond the scope of the data—perhaps about the effect of the learning environment on certain students, how students being pulled out of class for a sporting event the previous day impacted the results, or how some students have support at home while others do not. While these are all potential factors in determining results, at this stage, your collaborative team should focus on how the *data* reflect the efforts of your team during the unit's instruction.

Directions: With your collaborative team, answer the following questions after completing the end-of-unit assessment.

Team: _____   **Teacher:** _____

1. What went well overall in the unit?

2. What do the common end-of-unit data reveal about student performance on the essential learning standards with which students did well?

3. How did each teacher achieve these student successes?

4. What do the common end-of-unit data reveal about student performance on the essential learning standards with which students did not do well?

5. How well did we provide feedback to our students during the unit on the learning standards with which students struggled?

6. What elements of the unit instruction will we need to improve for future use?

7. What impact do the results from this unit have on instruction during our next unit based on our team's reflections? (RTI Tier 1)

8. How will our team provide for student re-engagement around essential learning standards that need more time, support, and focus for student learning? (RTI Tier 2 targeted team response)

Figure 3.9: Collaborative team data-analysis protocol.

Visit **go.solution-tree.com/mathematicsatwork** to download a reproducible version of this figure.

Collaboratively Scoring the Common Assessment Instrument

Scoring student work together will provide insight into other team members' mathematical thinking and allow you to consider multiple ways students and teachers represent their thinking. You can use the team activity in figure 3.10 to facilitate a collaborative scoring conversation.

Directions: Collect sample student work from your most recent common end-of-unit assessment. Make copies so that each team member has a complete set of student test samples to score, removing student names and numbering them for easy reference. Each team member should score the examples individually and make notes justifying his or her scoring decisions. Next, share your scoring and notes.

1. Did everyone score the student work for each mathematics task the same?

2. What reasons do team members have for their scores for the student work being different?

3. Develop a consensus score for each sample item. Ensure the score is based on mathematical understandings and appropriate problem-solving strategies as demonstrated by the student.

4. What strategies did you use to agree as a team on a common score for each task on the test?

5. Organize your reflections on the students' mathematical thinking. What are the patterns of thinking most prevalent for the tasks? What are the misconceptions that stood out? Did students use strategies on the test questions that were surprising?

Figure 3.10: Collaborative team scoring and calibration activity.

Visit **go.solution-tree.com/mathematicsatwork** to download a reproducible version of this figure.

After your collaborative team has analyzed the student work, each teacher then selects three examples of student work—one low, one middle, and one high score. Place the student name in the analysis of student work tool in figure 3.11, and complete the table.

Student Name	Description of Student Work	Student Strengths	Student Misconceptions	Future Work With Student
Implications for instruction:				

Figure 3.11: Analysis of student work tool.

Visit **go.solution-tree.com/mathematicsatwork** to download a reproducible version of this figure.

Fidelity occurs when your team meets to score student work fairly and consistently, so your team should practice common scoring of the same tasks at the end of every unit for the course throughout the year. For common assessment data to have meaning during your team analysis, students should receive similar scores from all team members (within one or two points for the entire end-of-unit exam).

Since time to collaborate is precious, figure 3.12 (page 164) provides a sample of an essential learning standards tracking tool that teacher teams can use for collecting after-the-unit assessment data. The example is from a seventh- and eighth-grade algebra 1 team whose members do not have a common planning time because of the grade-level difference. In order to still collaborate and share information, members use a common Google Doc to share their thoughts. When the grade-level algebra 1 teams meet, they can reference feedback from the other grade level and move forward.

Assessment	Main Areas of Student Confusion	Reengagement Strategies for Missed Concepts	Modifications to Our Lessons for Next Year	Adam's Median Test Score Period 1	Amy's Median Test Score Period 2, Period 3	David's Median Test Score Period 1, Period 5	Kelly's Median Test Score Period 9
Unit 0 (Grade 7 only): Geometry				92	90/92		91
Unit 1: Connecting patterns and functions				91	77/89	91/90	90
Unit 2: Linear functions				96	87.5/97	92/89	93
Unit 3: Modeling linear data	Students had a hard time understanding the relationship between a scatterplot and its residual plot. –DL			88	81.5/84	88.5/90	85
Unit 4: Modeling linear data	Students had a very hard time understanding that a frequency table was necessary to determine an association. –KM Agreed. –DL Students did not know they needed to create a relative frequency table in order to determine whether or not there is an association between categorical variables as evidenced by Quiz U4 L1–L3. After a day of review, students showed a significant improvement in understanding on the test. –AL Agreed. –AR Quiz average: 68% (AR)		Add a day to go over Quiz U4 L1–L3 on day 5 of the unit followed by a separate review day for the test.	92	90/93	88/92	

Figure 3.12: Sample assessment snapshot.

▨ = Assessment not given.

Source: Adapted with permission from Aptakisic Junior High, Buffalo Grove, Illinois.

Figure 3.13 is an essential learning standard tracking tool that teacher teams can use for more detailed data collection after the end of each assessment. These are the data the teachers would bring to the team conversation already prepared. The example shown in figure 3.13 is from a sixth-grade team unit 1 assessment. Teachers gathered the evidence individually and brought the completed form and student work to their collaborative time immediately following the unit assessment.

The team chose to track the number of students proficient on each essential standard (of various grain sizes) to identify strengths within the unit and then plan re-engagement strategies for concepts that students consistently missed. In areas where the misconceptions were minor, collaborative teams planned upcoming lessons and used scaffolding to address common misconceptions.

Specific strategies included using warm-ups that covered both the prerequisite knowledge for the lesson and previous unit misconceptions or adding review items to current homework to encourage continual practice of concepts that students mastered during the unit.

Unit 1: Add, Subtract, Multiply, and Divide Rational Numbers		**Teacher**			
		1	**2**	**3**	**4**
Content standard cluster: Apply and extend previous understandings of multiplication and division to divide fractions by fractions.		61%	45%	67%	70%
6.NS.1	Interpret and compute quotients of fractions, and solve word problems involving division of fractions by fractions.	61%	45%	67%	70%
Content standard cluster: Compute fluently with multidigit numbers and find common factors and multiples.		78%	61%	69%	79%
6.NS.2	Fluently divide multidigit numbers using the standard algorithm.	78%	79%	76%	89%
6.NS.3	Fluently add, subtract, multiply, and divide multidigit decimals using the standard algorithm for each operation.	77%	50%	63%	71%
Content standard cluster: Compute fluently with multi-digit numbers and find common factors and multiples.		77%	37%	37%	58%
6.NS.4	Find the greatest common factor of two whole numbers less than or equal to 100 and the least common multiple of two whole numbers less than or equal to 12. Use the distributive property to express a sum of two whole numbers 1–100 with a common factor as a multiple of a sum of two whole numbers with no common factor.	77%	37%	37%	58%
Content standard cluster: Solve real-world and mathematical problems involving area, surface area, and volume.		61%	38%	70%	75%
6.G.2	Apply the formulas $V = lwh$ and $V = Bh$ to find volumes of right rectangular prisms with fractional edge lengths in the context of solving real-world and mathematical problems.	61%	38%	70%	75%

Source for standards: NGA & CCSSO, 2010, pp. 42, 44, 45.

Figure 3.13: Unit 1 content standard cluster tracker by teacher.

Visit **go.solution-tree.com/mathematicsatwork** to download a reproducible version of this figure.

Teachers complete this analysis individually, and then, when they meet to review the results, they dig deeper into the student work to analyze student misconceptions. The sixth-grade team, using the example in figure 3.13, identified two standards (6.NS.1 and 6.NS.3) that challenged students. In this case, the team prefers to organize its analysis into bigger standard categories—around the CCSS mathematics content standard clusters for each essential standard. During the year, the teachers measure student progress in their mastery of these content standard clusters as they unfold in each unit. Upon reviewing student work on the test, the team was able to create re-engagement strategies to target student misconceptions.

Team members also noted that some teachers had better strengths with particular content. Teacher four crafted a re-engagement lesson for 6.G.2. Teacher one crafted a lesson for 6.NS.3, and the team shared the work for 6.NS.2. Then, they created a schedule for homeroom or intervention time (after school for some students) and posted the schedule in their classrooms so all students could attend the targeted Tier 2 intervention, even though it might not be with their assigned teacher.

One teacher was new to the course, and her students were not performing as well. The team was able to have an open conversation based on the data to improve the teacher's understanding of the progression of mathematical understanding for the unit. Some of the best conversations for student understanding occur when you discuss and review student work and patterns of errors. As teachers, you are sometimes limited to your perspective on how to teach a concept. The power of your team to open up new perspectives as you re-engage students in multiple solution pathways is of great benefit.

Setting Team Goals for Improvement

In order for you and your collaborative team to develop a cycle of continuous improvement, you need to set and reflect on collaborative team goals on a unit-by-unit basis. These goals have the same structure as the goals you use with students.

Consider the structure of clear and manageable team goals using the SMART acronym (O'Neill & Conzemius, 2006).

- **Strategic and specific:** Our goals are aligned to targeted essential learning standards.
- **Measureable:** We know when we achieve the essential learning standards.
- **Attainable:** We can achieve the goals through collaborative team efforts.
- **Results oriented:** We clearly define the outcome.
- **Time bound:** We have a specified time within which to achieve the goals.

Your team should set SMART goals (proficiency goals) for each of the unit's learning standards as you use the assessment results to prepare for the next unit. You can use a form such as the one available at **go.solution-tree.com/mathematicsatwork**. Your team can work on developing manageable short-term and long-term goals as well as orchestrating celebrations as students achieve the essential learning standards.

As an example, consider the data from the three essential learning standards (CCSS mathematics content standard clusters) for the unit shown in figure 3.13 (page 165). This team had set SMART goals of 75 percent proficiency for all students in the course. Team members should celebrate meeting this short-term goal for the content standard cluster, *Compute fluently with multidigit numbers and find common factors and multiples* (6.NS; NGA & CCSSO, 2010, p. 42). However, they need to plan for student re-engagement in the other two essential learning standards or clusters, based on the results of the assessment data.

You can use the reflection questions in figure 3.14 to analyze your team SMART goal performance on the assessment and your plans for instruction moving forward during the next unit.

Directions: Review your short-term SMART goals for the unit's essential learning standards.

1. How should your team respond to the success demonstrated in the data results (such as figure 3.13, on page 165)?

2. In what ways does your team define its expectation of student success prior to implementing the next unit?

3. How will your team respond to the students who did not achieve proficiency in this unit?

4. What SMART goals has your team set for the essential learning standards in the next unit? Why?

Figure 3.14: Collaborative team SMART goal-setting reflections.

Visit **go.solution-tree.com/mathematicsatwork** to download a reproducible version of this figure

In setting unit-by-unit student performance goals, your collaborative team needs to keep in mind that the focus should remain on student achievement—how can teachers help students grow in their mathematical understanding along all of the learning standards for the unit? Teachers should use student assessment data to consider the revision of both instructional and assessment practices within the collaborative team.

Figure 3.15 (page 168) shows a sample SMART goal review from a seventh-grade math team.

> **2014–2015 Student Learning Goals, Common Formative Assessment**
>
> **Team:** Seventh-grade mathematics
>
> **Team leader:** Suzanne Wright
>
> **Team members:** Correne Phillips, Corinne Howe, Susan Willis, and Katie Stroh
>
> **Student learning goal for semester one:** We want a minimum of 96 percent of all students who take seventh-grade mathematics to achieve a grade of C or higher.
>
> **Why did your team choose this goal?** Since our final exam will be changing, and the curriculum is changing, we felt we could focus on the success of our students in a different way this year.
>
> **Data from last year (for comparison) or data that suggest this needs improvement:** 78.2 percent of the students earned an A, B, or C.
>
> **Action steps team will take to meet the goal:** After each test, our team will track their progress of students' unit test grades and the proficiencies for each essential learning standard of the unit. At team meetings, we can discuss how things are going, and the teachers who have been more successful will share strategies that they are using in their class.
>
> **When will we complete these action steps?** We will complete these action steps after each unit test during the semester.
>
> **Results on student achievement goal for semester one:** 356 out of 433 of the students earned an A, B, or C; 82.2 percent of students passed with a C or higher.

Figure 3.15: Sample SMART goal expectations for a seventh-grade algebra 1 team.

On a larger scale, your course-based or grade-level team should consider several other types of longer-term SMART goals for the year. You should consider state achievement goals such as performance on state assessments, as well as local high school performance, especially investigating D/F rate performance after your students reach high school. For further study and greater depth of knowledge regarding the SMART goal process, see *Learning by Doing: A Handbook for Professional Learning Communities at Work* (DuFour et al., 2010) and *The Handbook for SMART School Teams: Revitalizing Best Practices for Collaboration* (Conzemius & O'Neill, 2014).

As a final point, consider the following questions in regard to your team's formative assessment process and end-of-unit assessment instruments.

- Did the tasks balance cognitive demand and rigor? Did students demonstrate this in their responses?

- In what ways did students demonstrate proficiency with the Mathematical Practices?

- How do the assessment instruments support student demonstration of depth of mathematical understanding of the essential learning standards?

Ask these questions throughout the reflective process as your team considers the unit tasks and assessment items, and also use them in the planning stages for the next unit. Student assessment performance should be a consideration for revisions in both instructional and assessment practices.

Your Team's Progress

It is helpful to diagnose your team's current reality and action after launching the unit. Ask each team member to individually assess your team on the tenth high-leverage team action using the status check tool in table 3.2. Discuss your perception of your team's progress on ensuring evidence-based adult goal setting and action for the next unit of study.

Table 3.2: After-the-Unit Status Check Tool for HLTA 10—Ensuring Evidence-Based Adult Goal Setting and Action for the Next Unit of Study

Directions: Discuss your perception of your team's progress on the tenth high-leverage team action—ensuring evidence-based adult goal setting and action for the next unit of study. Defend your reasoning.			
Stage I: Pre-Initiating	**Stage II: Initiating**	**Stage III: Developing**	**Stage IV: Sustaining**
We do not set team proficiency targets for each essential learning standard.	We consider end-of-unit assessment results, but we do not set proficiency targets for each essential learning standard.	We independently use student end-of-unit assessment results to determine if proficiency targets were achieved.	We collaboratively use student end-of unit assessment results to determine if proficiency targets were achieved.
We do not know the end-of-unit results of other team members.	We provide some constructive feedback to each other at the end of a unit if asked.	We score some assessments together and calibrate for scoring accuracy.	We score all assessments together and calibrate for scoring accuracy.
We do not analyze end-of-unit results.	We analyze end-of-unit results but this does not influence our planning for the next unit.	We carefully and independently consider how end-of-unit results impact our planning for the next unit.	We collaboratively and carefully consider how end-of-unit results impact our team planning for the next unit.

Visit **go.solution-tree.com/mathematicsatwork** to download a reproducible version of this table.

After the unit ends, you and your team should have planned, implemented, and reflected on the instructional unit, the feedback you provided to students during the unit, the assessment instruments you designed to gauge student understanding, and both the student and adult SMART goals or proficiency targets set for that unit. This is a lot to focus on, so you will need to set your team's priorities.

Setting Your After-the-Unit Priorities for Team Action

As part of a PLC culture, your collaborative teamwork is never over, even after the unit ends. You have worked hard to pursue the two high-leverage team actions we outlined in this chapter.

HLTA 9.   Ensuring evidence-based student goal setting and action for the next unit of study

HLTA 10. Ensuring evidence-based adult goal setting and action for the next unit of study

As a team, reflect together on the stage you identified for each of these team actions. Based on the results, what should be your team's priority? You can use figure 3.16 to focus your time and energy on actions that are most urgent after the unit ends. Focus on these few tasks, and make them matter at a deep level of implementation.

Directions: Identify the stage you rated your team for each of the two high-leverage team actions, and provide a brief rationale.

9. Ensuring evidence-based student goal setting and action for the next unit of study

Stage I: Pre-Initiating   Stage II: Initiating   Stage III: Developing   Stage IV: Sustaining

Reason: _____

10. Ensuring evidence-based adult goal setting and action for the next unit of study

Stage I: Pre-Initiating   Stage II: Initiating   Stage III: Developing   Stage IV: Sustaining

Reason: _____

With your collaborative team, respond to the red light, yellow light, and green light prompts for the high-leverage team actions that you and your team believe are most urgent to focus on.

Red light: Indicate one activity you will stop doing that limits effective implementation of the high-leverage team actions.

Yellow light: Indicate one activity you will continue to do to be effective with the high-leverage team actions.

Green light: Indicate one activity you will begin to do immediately to become more effective with the high-leverage team actions.

Figure 3.16: Setting your collaborative team's after-the-unit priorities.

Visit **go.solution-tree.com/mathematicsatwork** to download a reproducible version of this figure.

Your after-the-unit work consisted of (1) ensuring that each student receives effective feedback that is fair, accurate, specific, and timely and that students use the common assessment as a formative learning opportunity; and (2) ensuring consistent scoring for the common assessment instrument while ensuring students and adults use the common assessment results to identify achievement of proficiency targets and actions for the next unit of study and beyond. These pursuits address steps four and five of the PLC teaching-assessing-learning cycle (see figure 3.1, page 141).

The effective implementation of step five (your collaborative teams using ongoing assessment feedback to improve instruction) leads you to cycle back to step one and begin the process again with the next unit, setting new short-term goals and shifting focus to the next unit of instruction. The cycle is a continuous-improvement process necessary for effective mathematics teaching, assessing, and learning.

What is wonderful about the process of teaching and learning is that both students and teachers improve their understanding of mathematical content, Mathematical Practices, and role in the formative assessment process.

This chapter completes your investigation of the PLC teaching-assessing-learning cycle and the ten high-leverage team actions that deliver on the promise of improved student achievement.

Taking Your Next Steps

So now what? You and your collaborative team have moved through the stages of the PLC teaching-assessing-learning cycle and should now be ready to start the process again with the next unit. Some of the considerations from this handbook relative to work with your instructional unit include:

- Was the size of the unit manageable within the teaching-assessing-learning cycle?
- How did your team discussion of essential learning standards help you support student understanding?
- How did the design of the mathematical tasks and assessment instruments work? Were they aligned?
- How did the unit formative assessment plan fit with the end-of-unit assessment?

Figure E.1 (pages 174–175) provides a final summative evaluation your team can use at the beginning or the end of the school year to identify your current progress on each of the high-leverage team actions. Celebrate your strengths and prioritize your areas for continued growth.

Reflect on where your team falls along the continua for each high-leverage team action. The process of collaboration capitalizes on the fact that you come together to use diverse experiences and knowledge to create a whole that is larger than the sum of the parts. Your effective collaboration around these actions is *the* solution to sustained professional learning—an ongoing and never-ending process of teacher growth necessary to meet the expectations of the Common Core and beyond.

The National Board for Professional Teaching Standards (2010) states it like this:

> Seeing themselves as partners with other teachers, they are dedicated to improving the profession. They care about the quality of teaching in their schools, and, to this end, their collaboration with colleagues is continuous and explicit. They recognize that collaborating in a professional learning community contributes to their own professional growth, as well as to the growth of their peers, for the benefit of student learning. Teachers promote the ideal that working collaboratively increases knowledge, reflection, and quality of practice and benefits the instructional program. (p. 75)

The new paradigm for the professional development of mathematics teachers requires an understanding that the knowledge capacity of every teacher matters. More importantly, however, is that every teacher *acts* on that knowledge and transfers the professional development that he or she receives into his or her daily classroom practice.

Assessing Your High-Leverage Team Actions

Directions: Rate your team on a scale of 1 (low) to 6 (high) for your current implementation of each of the ten high-leverage team actions.

Before the Unit (Step 1 of the Cycle)

HLTA 1.   We agree on the expectations and intent of the common essential learning standards and Mathematical Practices for the unit.

Rating: _____

Reason: _____

_____

HLTA 2.   We identify and discuss student use of higher-level-cognitive-demand mathematical tasks as part of the instruction during the unit.

Rating: _____

Reason: _____

_____

HLTA 3.   We develop high-quality common assessment instruments for the unit.

Rating: _____

Reason: _____

_____

HLTA 4.   We develop accurate scoring rubrics and proficiency expectations for the common assessment instruments.

Rating: _____

Reason: _____

_____

HLTA 5.   We plan and use common homework assignments.

Rating: _____

Reason: _____

_____

During the Unit (Steps 2 and 3 of the Cycle)

HLTA 6.   We develop student proficiency in each Mathematical Practice through in-class, higher-level-cognitive-demand mathematical tasks.

Rating: _____

Reason: _____

_____

HLTA 7.   We use in-class formative assessment processes effectively.

Rating: _____

Reason: _____

_____

HLTA 8.   We use a lesson-design process for lesson planning and collective team inquiry.

Rating: _____

Reason: _____

_____

After the Unit (Steps 4 and 5 of the Cycle)

HLTA 9. We ensure evidence-based student goal setting and action for the next unit of study.

  Rating: _____

  Reason: _____

  _____

HLTA 10. We ensure evidence-based adult goal setting and action for the next unit of study.

  Rating: _____

  Reason: _____

  _____

Setting Your Collaborative Team Monitoring Priorities

Directions: Review your ratings for the ten high-leverage team actions under the three categories of essential PLC team commitments.

For each category, list your top two or three specific areas for improvement. Based on your knowledge of the team, what should be your focus for adult growth and improvement and knowledge capacity building? Be specific and use your ratings to inform your choices. Also examine your current progress on team-level SMART goals.

  1. PLC Teacher Team Agreements for Teaching and Learning Before the Unit Begins

  2. PLC Teacher Team Agreements for Teaching and Learning During the Unit

  3. PLC Teacher Team Agreements for Teaching and Learning After the Unit Ends

  4. PLC Teacher Team Agreements for SMART Goals

What data targets beckon for improved student achievement in each course or grade level for next year? Consider local (unit by unit), state (CCSS mathematics or otherwise), and national data-improvement targets.

Figure E.1: Tool for assessing your actions and setting your team priorities.

Visit **go.solution-tree.com/mathematicsatwork** to download a reproducible version of this figure.

APPENDIX A

Standards for Mathematical Practice

Source: NGA & CCSSO, 2010, pp. 6–8. © Copyright 2010. National Governors Association Center for Best Practices and Council of Chief State School Officers. All rights reserved.

The Standards for Mathematical Practice describe varieties of expertise that mathematics educators at all levels should seek to develop in their students. These practices rest on important "processes and proficiencies" with longstanding importance in mathematics education. The first of these are the NCTM process standards of problem solving, reasoning and proof, communication, representation, and connections. The second are the strands of mathematical proficiency specified in the National Research Council's report *Adding It Up:* adaptive reasoning, strategic competence, conceptual understanding (comprehension of mathematical concepts, operations and relations), procedural fluency (skill in carrying out procedures flexibly, accurately, efficiently and appropriately), and productive disposition (habitual inclination to see mathematics as sensible, useful, and worthwhile, coupled with a belief in diligence and one's own efficacy).

1. **Make sense of problems and persevere in solving them.** Mathematically proficient students start by explaining to themselves the meaning of a problem and looking for entry points to its solution. They analyze givens, constraints, relationships, and goals. They make conjectures about the form and meaning of the solution and plan a solution pathway rather than simply jumping into a solution attempt. They consider analogous problems, and try special cases and simpler forms of the original problem in order to gain insight into its solution. They monitor and evaluate their progress and change course if necessary. Older students might, depending on the context of the problem, transform algebraic expressions or change the viewing window on their graphing calculator to get the information they need. Mathematically proficient students can explain correspondences between equations, verbal descriptions, tables, and graphs or draw diagrams of important features and relationships, graph data, and search for regularity or trends. Younger students might rely on using concrete objects or pictures to help conceptualize and solve a problem. Mathematically proficient students check their answers to problems using a different method, and they continually ask themselves, "Does this make sense?" They can understand the approaches of others to solving complex problems and identify correspondences between different approaches.

2. **Reason abstractly and quantitatively.** Mathematically proficient students make sense of quantities and their relationships in problem situations. They bring two complementary abilities to bear on problems involving quantitative relationships: the ability to decontextualize—to abstract a given situation and represent it symbolically and manipulate the representing symbols as if they have a life of their own, without necessarily attending to their referents—and the ability to contextualize, to pause as needed during the manipulation process in order to probe into the referents for the symbols involved. Quantitative reasoning entails habits of creating a coherent representation of the problem at hand; considering the units involved; attending to the meaning of quantities, not just how to compute them; and knowing and flexibly using different properties of operations and objects.

3. Construct viable arguments and critique the reasoning of others. Mathematically proficient students understand and use stated assumptions, definitions, and previously established results in constructing arguments. They make conjectures and build a logical progression of statements to explore the truth of their conjectures. They are able to analyze situations by breaking them into cases, and can recognize and use counterexamples. They justify their conclusions, communicate them to others, and respond to the arguments of others. They reason inductively about data, making plausible arguments that take into account the context from which the data arose. Mathematically proficient students are also able to compare the effectiveness of two plausible arguments, distinguish correct logic or reasoning from that which is flawed, and—if there is a flaw in an argument—explain what it is. Elementary students can construct arguments using concrete referents such as objects, drawings, diagrams, and actions. Such arguments can make sense and be correct, even though they are not generalized or made formal until later grades. Later, students learn to determine domains to which an argument applies. Students at all grades can listen or read the arguments of others, decide whether they make sense, and ask useful questions to clarify or improve the arguments.

4. Model with mathematics. Mathematically proficient students can apply the mathematics they know to solve problems arising in everyday life, society, and the workplace. In early grades, this might be as simple as writing an addition equation to describe a situation. In middle grades, a student might apply proportional reasoning to plan a school event or analyze a problem in the community. By high school, a student might use geometry to solve a design problem or use a function to describe how one quantity of interest depends on another. Mathematically proficient students who can apply what they know are comfortable making assumptions and approximations to simplify a complicated situation, realizing that these may need revision later. They are able to identify important quantities in a practical situation and map their relationships using such tools as diagrams, two-way tables, graphs, flowcharts and formulas. They can analyze those relationships mathematically to draw conclusions. They routinely interpret their mathematical results in the context of the situation and reflect on whether the results make sense, possibly improving the model if it has not served its purpose.

5. Use appropriate tools strategically. Mathematically proficient students consider the available tools when solving a mathematical problem. These tools might include pencil and paper, concrete models, a ruler, a protractor, a calculator, a spreadsheet, a computer algebra system, a statistical package, or dynamic geometry software. Proficient students are sufficiently familiar with tools appropriate for their grade or course to make sound decisions about when each of these tools might be helpful, recognizing both the insight to be gained and their limitations. For example, mathematically proficient high school students analyze graphs of functions and solutions generated using a graphing calculator. They detect possible errors by strategically using estimation and other mathematical knowledge. When making mathematical models, they know that technology can enable them to visualize the results of varying assumptions, explore consequences, and compare predictions with data. Mathematically proficient students at various grade levels are able to identify relevant external mathematical resources, such as digital content located on a website, and use them to pose or solve problems. They are able to use technological tools to explore and deepen their understanding of concepts.

6. Attend to precision. Mathematically proficient students try to communicate precisely to others. They try to use clear definitions in discussion with others and in their own reasoning. They state the meaning of the symbols they choose, including using the equal sign consistently and appropriately. They are careful about specifying units of measure, and labeling axes to clarify the correspondence with quantities in a problem. They calculate accurately and efficiently, and express numerical answers with a degree of precision appropriate for the problem context. In the elementary grades, students give carefully formulated explanations to each other. By the time they reach high school they have learned to examine claims and make explicit use of definitions.

7. **Look for and make use of structure.** Mathematically proficient students look closely to discern a pattern or structure. Young students, for example, might notice that three and seven more is the same amount as seven and three more, or they may sort a collection of shapes according to how many sides the shapes have. Later, students will see 7×8 equals the well remembered $7 \times 5 + 7 \times 3$, in preparation for learning about the distributive property. In the expression $x^2 + 9x + 14$, older students can see the 14 as 2×7 and the 9 as $2 + 7$. They recognize the significance of an existing line in a geometric figure and can use the strategy of drawing an auxiliary line for solving problems. They also can step back for an overview and shift perspective. They can see complicated things, such as some algebraic expressions, as single objects or as being composed of several objects. For example, they can see $5 - 3(x - y)^2$ as 5 minus a positive number times a square and use that to realize that its value cannot be more than 5 for any real numbers x and y.

8. **Look for and express regularity in repeated reasoning.** Mathematically proficient students notice if calculations are repeated, and look both for general methods and for shortcuts. Upper elementary students might notice when dividing 25 by 11 that they are repeating the same calculations over and over again, and conclude they have a repeating decimal. By paying attention to the calculation of slope as they repeatedly check whether points are on the line through $(1, 2)$ with slope 3, middle school students might abstract the equation $(y - 2)/(x - 1) = 3$. Noticing the regularity in the way terms cancel when expanding $(x - 1)(x + 1)$, $(x - 1)(x^2 + x + 1)$, and $(x - 1)(x^3 + x^2 + x + 1)$ might lead them to the general formula for the sum of a geometric series. As they work to solve a problem, mathematically proficient students maintain oversight of the process, while attending to the details. They continually evaluate the reasonableness of their intermediate results.

Connecting the Standards for Mathematical Practice to the Standards for Mathematical Content

The Standards for Mathematical Practice describe ways in which developing student practitioners of the discipline of mathematics increasingly ought to engage with the subject matter as they grow in mathematical maturity and expertise throughout the elementary, middle and high school years. Designers of curricula, assessments, and professional development should all attend to the need to connect the mathematical practices to mathematical content in mathematics instruction.

The Standards for Mathematical Content are a balanced combination of procedure and understanding. Expectations that begin with the word "understand" are often especially good opportunities to connect the practices to the content. Students who lack understanding of a topic may rely on procedures too heavily. Without a flexible base from which to work, they may be less likely to consider analogous problems, represent problems coherently, justify conclusions, apply the mathematics to practical situations, use technology mindfully to work with the mathematics, explain the mathematics accurately to other students, step back for an overview, or deviate from a known procedure to find a shortcut. In short, a lack of understanding effectively prevents a student from engaging in the mathematical practices.

In this respect, those content standards which set an expectation of understanding are potential "points of intersection" between the Standards for Mathematical Content and the Standards for Mathematical Practice. These points of intersection are intended to be weighted toward central and generative concepts in the school mathematics curriculum that most merit the time, resources, innovative energies, and focus necessary to qualitatively improve the curriculum, instruction, assessment, professional development, and student achievement in mathematics.

Standards for Mathematical Practice Evidence Tool

Source: © Copyright 2013 by Mona Toncheff & Timothy D. Kanold. All rights reserved.
Source for Mathematical Practices: NGA & CCSSO, 2010, p. 6–8.

Mathematical Practice 1: "Make Sense of Problems and Persevere in Solving Them"

Students:

- Check intermediate answers, and change strategy if necessary
- Think about approaches to solving the problem before beginning
- Draw pictures or diagrams to represent given information
- Have the patience to complete multiple examples in trying to identity a solution
- Start by working a simpler problem
- Make a plan for solving the problem

In the classroom:

- Student teams or groups look at a variety of solution approaches and discuss their merits.
- Student teams or groups compare two different approaches to look for connections.
- Students discuss with their peers whether a particular answer is possible in a given situation and explain their thinking.
- Time is allotted for individual thinking and for sharing thoughts and ideas with a student peer.

Mathematical Practice 2: "Reason Abstractly and Quantitatively"

Students:

- Connect numbers and symbols with a concrete model
- Write and manipulate symbols to solve a problem
- Generalize a pattern (such as, "I see that $i^2 = -1$, $i^3 = -i$, $i^4 = 1$, so i^{257} must be _____ since the pattern is _____")
- Approximate the answer by reasoning quantitatively before doing the actual calculation

In the classroom:

- Students explain the meaning of a given expression or equation in terms of a situation to their peers.
- Students interpret an answer in the context of a problem and verify their reasoning with a peer.

- Students discuss with peers and choose the most likely answer to a given situation without actually doing any calculations.

Mathematical Practice 3: "Construct Viable Arguments and Critique the Reasoning of Others"

Students:

- Explain what is wrong in a process that has led to an incorrect answer
- Justify why an answer makes sense or is reasonable
- Justify each step of a proof argument
- Explain why they chose one possible answer over another for a given problem
- Use a specific example to test a general conjecture

In the classroom:

- Students explain their thinking process to those in their team or group.
- Students justify why they agree or disagree with an answer.
- Students discuss different ways to represent a pattern that they see and explain why their representation works to their peers.
- Students group objects that are similar and explain why they think so.
- Students debate with their peers whether arguments in a proof are mathematically reasonable.

Mathematical Practice 4: "Model With Mathematics"

Students:

- Model a real-world scenario with a mathematical representation
- Use a variety of modalities to represent different scenarios
- Represent a story with concrete objects
- Draw a picture to illustrate a problem
- Analyze data to determine whether a conclusion is reasonable

In the classroom:

- Students work in pairs or groups to design a mathematical model for a given situation.
- Student teams and groups act out a problem scenario or use manipulatives to demonstrate it.
- Students brainstorm the mathematics they have previously learned that they need to represent and solve the situation.

Mathematical Practice 5: "Use Appropriate Tools Strategically"

Students:

- Use algebra tiles to investigate how a pattern develops

- Use compass rulers, protractors, and patty paper to investigate the properties of a geometric figure
- Use graphs and other visual representations to help determine rules about expressions and equations
- Apply relevant formulas correctly to a situation (for example, area, volume, quadratic, and distance)
- Use geometric software to test conjectures
- Use graphing calculators for tasks that are inappropriate to do by hand (like graphing complicated functions, calculating intersections of two graphs, finding a least squares regression line, and so on)

In the classroom:

- Teachers demonstrate appropriate and inappropriate uses of a variety of tools.
- Tools are available for students to use when needed.
- Students share how they chose and used various tools in their solutions.

Mathematical Practice 6: "Attend to Precision"

Students:

- Specify units of measure in their answers
- Define symbols and variables that they are using
- Use vocabulary appropriately
- Measure accurately
- Calculate precisely
- Label graphs and tables correctly
- Recognize and discard extraneous solutions

In the classroom:

- Teacher models appropriate use of mathematical vocabulary.
- Teacher models calculating with units throughout a problem, not just adding them on at the end.
- Student teams or groups discuss appropriate accuracy of answers.
- Student teams or groups discuss differences between a sketch and a graph.
- During student team or group presentations, other class members and the teacher help to clarify language and explanations.

Mathematical Practice 7: "Look for and Make Use of Structure"

Students:

- Look for efficient solution strategies

- Recognize common properties of geometric figures and use them to find answers

- Recognize how to use properties of numbers or equations to simplify a solution

- Identify the sign or the magnitude that a particular answer must have by recognizing a structure in the calculation (for example, answer must be negative or answer must be less than 1)

- Draw auxiliary lines on a geometric figure to reveal a characteristic

- Perform mental arithmetic by rearranging or separating terms to make the calculating easier (for example, in 97 + 105 = 100 + 102, moving 3 to the 97 to make 100)

In the classroom:

- Teacher demonstrates the structure in a complicated expression to help in simplifying it.

- Student teams or groups explain shortcut methods.

- Students outline the structure they've identified or recognized for the rest of the group.

Mathematical Practice 8: "Look for and Express Regularity in Repeated Reasoning"

Students:

- Use their experience in previous work to explain why an answer cannot be possible (such as, "The answer can't be more than 1 because you're dividing a smaller number by a larger one")

- Look for a way to express a shortcut to a solution

- Employ the repeated reasoning to skip steps in a solution

In the classroom:

- Teacher listens for aha moments and extends that thinking for the entire class.

- Student teams or groups discuss their observations, looking for common threads and general rules or properties.

APPENDIX C

Cognitive-Demand-Level Task-Analysis Guide

Source: Smith & Stein, 1998. Copyright 1998, National Council of Teachers of Mathematics. Used with permission.

Table C.1: Cognitive-Demand Levels of Mathematical Tasks

Lower-Level Cognitive Demand	Higher-Level Cognitive Demand
Memorization Tasks • These tasks involve reproducing previously learned facts, rules, formulae, or definitions to memory. • They cannot be solved using procedures because a procedure does not exist or because the time frame in which the task is being completed is too short to use the procedure. • They are not ambiguous; such tasks involve exact reproduction of previously seen material and what is to be reproduced is clearly and directly stated. • They have no connection to the concepts or meaning that underlie the facts, rules, formulae, or definitions being learned or reproduced.	**Procedures With Connections Tasks** • These procedures focus students' attention on the use of procedures for the purpose of developing deeper levels of understanding of mathematical concepts and ideas. • They suggest pathways to follow (explicitly or implicitly) that are broad general procedures that have close connections to underlying conceptual ideas as opposed to narrow algorithms that are opaque with respect to underlying concepts. • They usually are represented in multiple ways (for example, visual diagrams, manipulatives, symbols, or problem situations). They require some degree of cognitive effort. Although general procedures may be followed, they cannot be followed mindlessly. Students need to engage with the conceptual ideas that underlie the procedures in order to successfully complete the task and develop understanding.
Procedures Without Connections Tasks • These procedures are algorithmic. Use of the procedure is either specifically called for, or its use is evident based on prior instruction, experience, or placement of the task. • They require limited cognitive demand for successful completion. There is little ambiguity about what needs to be done and how to do it. • They have no connection to the concepts or meaning that underlie the procedure being used. • They are focused on producing correct answers rather than developing mathematical understanding. • They require no explanations or have explanations that focus solely on describing the procedure used.	**Doing Mathematics Tasks** • Doing mathematics tasks requires complex and no algorithmic thinking (for example, the task, instructions, or examples do not explicitly suggest a predictable, well-rehearsed approach or pathway). • It requires students to explore and understand the nature of mathematical concepts, processes, or relationships. • It demands self-monitoring or self-regulation of one's own cognitive processes. • It requires students to access relevant knowledge and experiences and make appropriate use of them in working through the task. • It requires students to analyze the task and actively examine task constraints that may limit possible solution strategies and solutions. • It requires considerable cognitive effort and may involve some level of anxiety for the student due to the unpredictable nature of the required solution process.

Sources for Higher-Level-Cognitive-Demand Tasks

Common Core Conversation

www.commoncoreconversation.com/math-resources.html

Common Core Conversation is a collection of more than fifty free website resources for the Common Core State Standards in mathematics and ELA.

EngageNY Mathematics

www.engageny.org/mathematics

The site features curriculum modules from the state of New York that include sample assessment tasks, deep resources, and exemplars for grades preK–12.

Howard County Public School System Secondary Mathematics Common Core

https://secondarymathcommoncore.wikispaces.hcpss.org

This site is a sample wiki for a district K–12 mathematics curriculum.

Illustrative Mathematics

www.illustrativemathematics.org

The main goal of this project is to provide guidance to states, assessment consortia, testing companies, and curriculum developers by illustrating the range and types of mathematical work that students will experience upon implementation of the Common Core State Standards for mathematics.

Inside Mathematics

www.insidemathematics.org/index.php/common-core-standards

This site provides classroom videos and lesson examples to illustrate the Mathematical Practices.

Mathematics Assessment Project

http://map.mathshell.org/materials/index.php

The Mathematics Assessment Project (MAP) aims to bring to life the Common Core State Standards in a way that will help teachers and their students turn their aspirations for achieving them into classroom realities. MAP is a collaboration between the University of California at Berkeley; the Shell Centre team at the University of Nottingham; and the Silicon Valley Mathematics Initiative (MARS).

Mathematics Vision Project

www.mathematicsvisionproject.org

The site features integrated high school curriculum modules that include mathematics performance tasks and video modules connected to Khan Academy.

National Council of Supervisors of Mathematics

www.mathedleadership.org/ccss/itp/index.html

This site features collections of K–12 mathematical tasks for illustrating the Standards for Mathematical Practice. The website includes best-selling books, numerous journal articles, and insights into the teaching and learning of mathematics.

National Council of Teachers of Mathematics Illuminations

http://illuminations.nctm.org

This site provides standards-based resources that improve the teaching and learning of mathematics for all students. The materials illuminate the vision for school mathematics set forth in NCTM's *Principles and Standards for School Mathematics, Curriculum Focal Points for Prekindergarten Through Grade 8 Mathematics*, and *Focus in High School Mathematics: Reasoning and Sense Making*.

National Science Digital Library

http://nsdl.org/commcore/math

The National Science Digital Library (NSDL) contains digital learning objects and tasks that are related to specific Common Core State Standards for mathematics.

Partnership for Assessment of Readiness for College and Careers Task Prototypes and New Sample Items for Mathematics

www.parcconline.org/samples/math

This page contains sample web-based practice assessment tasks (released items) for your use.

Smarter Balanced Assessment Consortium Sample Items and Performance Tasks

www.smarterbalanced.org/sample-items-and-performance-tasks

This site contains sample higher-level-cognitive-demand tasks and online test-taking and performance-assessment tasks (released items) for your use in class.

Virginia Department of Education

www.doe.virginia.gov/instruction/mathematics/index.shtml

This site contains mathematical tasks and teacher team materials to use with the tasks for grades 3–12.

Visit **go.solution-tree.com/mathematicsatwork** for continued updates on this resource list.

How the Mathematics at Work High-Leverage Team Actions Support the NCTM *Principles to Actions: Ensuring Mathematical Success for All*

The *Beyond the Common Core: A Handbook for Mathematics in a PLC at Work* series and the Mathematics at Work process include ten high-leverage team actions teachers should pursue collaboratively every day, in every unit, and every year. The goals of these actions are to eliminate inequities, inconsistencies, and lack of coherence so the focus is on teachers' expectations, instructional practices, assessment practices, and responses to student-demonstrated learning. Therefore, the Mathematics at Work process provides support for NCTM's Guiding Practices for School Mathematics as outlined in the 2014 publication *Principles to Actions: Ensuring Mathematical Success for All* (p. 5). Those principles are:

- **Curriculum principle**—An excellent mathematics program includes a curriculum that develops important mathematics along coherent learning progressions and develops connections among areas of mathematical study and between mathematics and the real world.

- **Professionalism principle**—In an excellent mathematics program, educators hold themselves and their colleagues accountable for the mathematical success of every student and for personal and collective professional growth toward effective teaching and learning of mathematics.

- **Teaching and learning principle**—An excellent mathematics program requires effective teaching that engages students in meaningful learning through individual and collaborative experiences that promote their ability to make sense of mathematical ideas and reason mathematically.

- **Assessment principle**—An excellent mathematics program ensures that assessment is an integral part of instruction, provides evidence of proficiency with important mathematics content and practices, includes a variety of strategies and data sources, and informs feedback to students, instructional decisions, and program improvement.

- **Access and equity principle**—An excellent mathematics program requires that all students have access to a high-quality mathematics curriculum, effective teaching and learning, high expectations, and the support and resources needed to maximize their learning potential.

- **Tools and technology principle**—An excellent mathematics program integrates the use of mathematical tools and technology as essential resources to help students learn and make sense of mathematical ideas, reason mathematically, and communicate their ideas.

Table E.1 (pages 190–191) shows how the HLTAs support NCTM's principles.

Table E.1: The HLTAs and NCTM *Principles to Actions*

Mathematics at Work High-Leverage Team Actions	NCTM's Guiding Practices for School Mathematics
HLTA 1. Making sense of the agreed-on essential learning standards (content and practices) and pacing What do we want all students in each grade level or course to know, understand, demonstrate, and be able to do? Procedures are in place that ensure teacher teams align the most effective mathematical tasks and instructional strategies to the content progression established in the overall unit plan components.	**Curriculum principle** **Professionalism principle.** The professionalism principle specifically calls for teachers to collaboratively examine and prioritize the mathematics content and Mathematical Practices that students are to learn. **Teaching and learning principle.** The teaching and learning principle establishes mathematics goals to focus learning. **Tools and technology principle**
HLTA 2. Identifying higher-level-cognitive-demand mathematical tasks Teacher teams choose mathematical tasks that represent a balance of higher- and lower-level cognitive demand for the essential learning standards of the unit of study.	**Teaching and learning principle.** Effective teaching and learning practices include implementing tasks that promote reasoning and problem solving and supporting productive struggle in learning mathematics. **Tools and technology principle**
HLTA 3. Developing common assessment instruments Develop, design, and create common end-of-unit assessments as a team before the unit begins based on high-quality design and test-evaluation tools. Ensure the assessment instruments are aligned with the instructional discussions and practices used during the unit and connected to all aspects of the essential learning standards for the unit.	**Assessment principle** **Professionalism principle.** The professionalism principle specifically calls for teachers to collaboratively develop and use common assessments. **Tools and technology principle** **Access and equity principle**
HLTA 4. Developing scoring rubrics and proficiency expectations for the common assessment instruments Design common scoring rubrics and assessment practices to align with expected student reasoning and proficiency for every essential learning standard of the unit.	**Assessment principle** **Access and equity principle**
HLTA 5. Planning and using common homework assignments Homework should be viewed as a daily opportunity for formative self-assessment and independent practice for students. Homework protocols include limiting the number of daily tasks, providing spaced practice, balancing cognitive-demand levels, providing all assignments to the students in advance of the unit, and carefully aligning the essential learning standards for the unit.	**Assessment principle** **Access and equity principle**

HLTA 6. Using higher-level-cognitive-demand mathematical tasks effectively	**Teaching and learning principle.** Effective teaching and learning practices include implementing tasks that promote reasoning and problem solving and supporting productive struggle in learning mathematics.
Teachers provide targeted and differentiated in-class support as students engage in mathematical processes and peer-to-peer discussions for learning by using higher-level-cognitive-demand tasks in every lesson.	**Tools and technology principle**
HLTA 7. Using in-class formative assessment processes effectively	**Assessment principle**
Teacher teams do deep planning for small-group discourse and peer-to-peer in-class formative assessment processes via meaningful, specific, and timely teacher feedback with subsequent student action. This requires much more than the diagnostic tool of checking for understanding. To be formative, students must receive feedback during class and take action on that feedback.	**Teaching and learning principle.** Effective teaching and learning practices include eliciting and using evidence of student thinking.
	Access and equity principle
Teachers intentionally use differentiated and targeted scaffolding and advancing Tier 1 RTI supports as students engage in higher-level-cognitive-demand tasks.	
HLTA 8. Using a lesson-design process for lesson planning and collective team inquiry	**Professionalism principle.** The professionalism principle specifically calls for teachers to collaboratively discuss, select, and implement common research-informed instructional strategies and plans.
Teachers ensure all lesson elements contain successful opportunities for student demonstration of understanding, with feedback and action on student learning.	**Teaching and learning principle.** All lesson designs should draw from the eight research-informed mathematics teaching practices.
Teachers actively engage in a teacher team–developed and team-designed lesson, observe teachers teaching the lesson, and debrief the lesson in order to learn from colleagues.	**Tools and technology principle**
HLTA 9. Ensuring evidence-based student goal setting and action for the next unit of study	**Teaching and learning principle.** Effective teaching and learning practices include eliciting and using evidence of student thinking.
Teachers and teacher teams require students to correct their errors and identify the essential learning standards that are strengths and weaknesses based on the results of the end-of-unit assessment.	**Assessment principle**
	Access and equity principle
Teachers work with students to complete and carry out a plan for improvement and action based on end-of-unit assessment results and outcomes for proficiency.	
HLTA 10. Ensuring evidence-based adult goal setting and action for the next unit of study	**Assessment principle**
Teachers and teacher teams score all end-of-unit assessments together and calibrate scoring to ensure accuracy and freedom from bias.	**Professionalism principle.** The professionalism principle specifically calls for teachers to collaboratively develop action plans that they can implement when students demonstrate that they have or have not attained the standards.
Teachers work together after the unit to determine if proficiency targets for students were achieved.	**Access and equity principle**
Teachers collaboratively and carefully consider how end-of-unit results are used to impact instruction and team planning for the next unit.	

References and Resources

Anderson, J. R., Reder, L. M., & Simon, H. A. (1995). *Applications and misapplications of cognitive psychology to mathematics education.* Unpublished paper, Carnegie Mellon University, Department of Psychology, Pittsburgh, PA. Accessed at http://act-r.psy.cmu.edu/papers/misapplied.html on October 1, 2014.

Black, P., & Wiliam, D. (2001). *Inside the black box: Raising standards through classroom assessment.* London: Assessment Group of the British Educational Research Association.

Boston, M. D., & Smith, M. S. (2009). Transforming secondary mathematics teaching: Increasing the cognitive demands of instructional tasks used in teachers' classrooms. *Journal for Research in Mathematics Education, 40*(2), 119–156.

Butler, R. (1988). Enhancing and undermining intrinsic motivation: The effects of task-involving and ego-involving evaluation on interest and performance. *British Journal of Educational Psychology, 58*(1), 1–14.

Chappuis, S., & Stiggins, R. J. (2002). Classroom assessment for learning. *Educational Leadership, 60*(1), 40–43.

Collins, J., & Hansen, M. T. (2011). *Great by choice: Uncertainty, chaos, and luck—Why some thrive despite them all.* New York: HarperCollins.

Common Core State Standards Initiative. (2014). *Key shifts in mathematics.* Accessed at www.corestandards.org/other-resources/key-shifts-in-mathematics on July 15, 2014.

Conzemius, A. E., & O'Neill, J. (2014). *The handbook for SMART school teams: Revitalizing best practices for collaboration* (2nd ed.). Bloomington, IN: Solution Tree Press.

Cooper, H. (2008a). *Effective homework assignments* (Research brief). Reston, VA: National Council of Teachers of Mathematics.

Cooper, H. (2008b). *Homework: What the research says* (Research brief). Reston, VA: National Council of Teachers of Mathematics.

Darling-Hammond, L. (2014). Testing to, and beyond, the Common Core. *Principal,* 8–12. Accessed at www.naesp.org/sites/default/files/Darling-Hammond_JF14.pdf on February 7, 2014.

Davies, A. (2007). Involving students in the classroom assessment process. In Douglas Reeves (Ed.), *Ahead of the curve: The power of assessment to transform teaching and learning* (pp. 31–57). Bloomington, IN: Solution Tree Press.

Dixon, J. K., & Tobias, J. M. (2013). The "whole" story: Understanding fraction computation. *Mathematics Teaching in the Middle School, 19*(3), 156–163.

DuFour, R., DuFour, R., & Eaker, R. (2008). *Revisiting Professional Learning Communities at Work: New insights for improving schools.* Bloomington, IN: Solution Tree Press.

DuFour, R., DuFour, R., Eaker, R., & Karhanek, G. (2010). *Raising the bar and closing the gap: Whatever it takes.* Bloomington, IN: Solution Tree Press.

DuFour, R., DuFour, R., Eaker, R., & Many, T. (2010). *Learning by doing: A handbook for Professional Learning Communities at Work* (2nd ed.). Bloomington, IN: Solution Tree Press.

Dweck, C. S. (2007). *Mindset: The new psychology of success.* New York: Ballantine Books.

EngageNY. (2013). *Algebra I, module 1.* Accessed at https://content.engageny.org/resource/algebra-i-module-1 on May 10, 2014.

Fisher, D., Frey, N., & Rothenberg, C. (2010). *Implementing RTI with English learners.* Bloomington, IN: Solution Tree Press.

Foster, D. (2008, November). *The real change agents: Building professional learning community.* Presentation at the California Math Conference. Accessed at www.noycefdn.org/documents/math/real changeagents.pdf on March 14, 2014.

Frayer, D. A., Fredrick, W. C., & Klausmeier, H. J. (1969). *A schema for testing the level of mastery.* Madison: Wisconsin Research and Development Center for Cognitive Learning.

Fullan, M. (2008). *The six secrets of change: What the best leaders do to help their organizations survive and thrive.* San Francisco: Jossey-Bass.

Gersten, R., Taylor, M. J., Keys, T. D., Rolfhus, E., & Newman-Gonchar, R. (2014). *Summary of research on the effectiveness of math professional development approaches.* (REL 2014–010). Washington, DC: National Center for Education Evaluation and Regional Assistance.

Gladwell, M. (2008). *Outliers: The story of success.* New York: Little, Brown.

Hattie, J. A. C. (2009). *Visible learning: A synthesis of over 800 meta-analyses relating to achievement.* New York: Routledge.

Hattie, J. A. C. (2012). *Visible learning for teachers: Maximizing impact on learning.* New York: Routledge.

Heflebower, T., Hoegh, J. K., Warrick, P., with Clemens, B., Hoback, M., & McInteer, M. (2014). *A school leader's guide to standards-based grading.* Bloomington, IN: Marzano Research Laboratory.

Herman, J., & Linn, R. (2013). *On the road to assessing deeper learning: The status of Smarter Balanced and PARCC assessment consortia* (CRESST Report 823). Los Angeles: University of California, National Center for Research on Evaluation, Standards, and Student Testing.

Hiebert, J. S., & Grouws, D. A. (2007). The effects of classroom mathematics teaching on students' learning. In F. K. Lester Jr. (Ed.), *Second handbook of research on mathematics teaching and learning: A project of the National Council of Teachers of Mathematics* (pp. 371–404). Charlotte, NC: Information Age.

Hiebert, J., & Stigler, J. W. (2000). A proposal for improving classroom teaching: Lessons from the TIMSS video study. *The Elementary School Journal, 101*(1), 3–20.

Jackson, K., Garrison, A., Wilson, J., Gibbons, L., & Shahan, E. (2013). Exploring relationships between setting up complex tasks and opportunities to learn in concluding whole-class discussions in middle-grades mathematics instruction. *Journal for Research in Mathematics Education, 44*(4), 646–682.

Johnson, D. W., & Johnson, R. T. (1999). Making cooperative learning work. *Theory Into Practice, 38*(2), 67–73.

Johnson, D. W., Johnson, R. T., & Holubec, E. J. (2008). *Cooperation in the classroom* (Rev. ed.). Edina, MN: Interaction Book.

Kagan, S. (1994). *Cooperative learning.* San Clemente, CA: Kagan.

Kagan, S., & Kagan, M. (2009). *Kagan cooperative learning.* San Clemente, CA: Kagan.

Kanold, T. D. (2011). *The five disciplines of PLC leaders.* Bloomington, IN: Solution Tree Press.

Kanold, T. D. (Ed.), Briars, D. J., Asturias, H., Foster, D., & Gale, M. A. (2013). *Common Core mathematics in a PLC at Work, grades 6–8.* Bloomington, IN: Solution Tree Press.

Kanold, T. D., Briars, D. J., & Fennell, F. (2012). *What principals need to know about teaching and learning mathematics.* Bloomington, IN: Solution Tree Press.

Kanold, T. D. (Ed.), Kanold, T. D., & Larson, M. R. (2012). *Common Core mathematics in a PLC at Work, leader's guide.* Bloomington, IN: Solution Tree Press.

Kanold, T. D. (Ed.), Larson, M. R., Fennell, F., Adams, T. L., Dixon, J. K., Kobett, B. M., & Wray, J. A. (2012a). *Common Core mathematics in a PLC at Work, grades K–2.* Bloomington, IN: Solution Tree Press.

Kanold, T. D. (Ed.), Larson, M., Fennell, F., Adams, T. L., Dixon, J. K., Kobett, B. M., & Wray, J. A. (2012b). *Common Core mathematics in a PLC at Work, grades 3–5.* Bloomington, IN: Solution Tree Press.

Kanold, T. D. (Ed.), Briars, D. J., Asturias, H., Foster, D., & Gale, M. A. (2012). *Common core mathematics in a PLC at Work, grades 6–8.* Bloomington, IN: Solution Tree Press.

Kanold, T. D. (Ed.), Zimmermann, G., Carter, J. A., Kanold, T. D., & Toncheff, M. (2012). *Common Core mathematics in a PLC at Work, high school.* Bloomington, IN: Solution Tree Press.

Kennedy, M. M. (2010). Attribution error and the quest for teacher quality. *Educational Researcher, 39*(8), 591–598.

Kilpatrick, J., Swafford, J., & Findell, B. (Eds.). (2001). *Adding it up: Helping children learn mathematics.* Washington, DC: National Academies Press.

Lappan, G., & Briars, D. (1995). How should mathematics be taught? In I. M. Carl (Ed.), *Prospects for school mathematics: Seventy-five years of progress* (pp. 131–156). Reston, VA: National Council of Teachers of Mathematics.

Larson, M. R. (2011). *Administrator's guide to interpreting the common core state standards to improve mathematics education.* Reston, VA: National Council of Teachers of Mathematics.

Marzano, R. J. (2007). *The art and science of teaching: A comprehensive framework for effective instruction.* Alexandria, VA: Association for Supervision and Curriculum Development.

Marzano, R. J. (2010). *Formative assessment & standards-based grading.* Bloomington, IN: Marzano Research.

Morris, A. K., Hiebert, J., & Spitzer, S. M. (2009). Mathematical knowledge for teaching in planning and evaluating instruction: What can preservice teachers learn? *Journal for Research in Mathematics Education, 40*(5), 491–529.

Mueller, C. M., & Dweck, C. S. (1998). Praise for intelligence can undermine children's motivation and performance. *Journal of Personality and Social Psychology, 75*(1), 33–52.

National Board for Professional Teaching Standards. (2010). *National Board for Teacher Certification: Mathematics standards for teachers of students ages 11–18+.* Arlington, VA: Author.

National Council of Teachers of Mathematics. (1991). *Professional standards for teaching mathematics.* Reston, VA: Author.

National Council of Teachers of Mathematics. (2007). *Mathematics teaching today: Improving practice, improving student learning.* Reston, VA: Author.

National Council of Teachers of Mathematics. (2012). *Fuel for thought.* Accessed at http://www.nctm.org /uploadedFiles/Journals_and_Books/Books/FHSM/RSM-Task/Fuel_for_Thought.pdf on May 10, 2014.

National Council of Teachers of Mathematics. (2014). *Principles to actions: Ensuring mathematical success for all.* Reston, VA: Author.

National Governors Association Center for Best Practices & Council of Chief State School Officers. (2010). *Common Core State Standards for mathematics.* Washington, DC: Authors. Accessed at www.corestandards.org/assets/CCSSI_Math%20Standards.pdf on February 7, 2014.

O'Neill, J., & Conzemius, A. (2006). *The power of SMART goals: Using goals to improve student learning.* Bloomington, IN: Solution Tree Press.

Partnership for Assessment of Readiness for College and Careers (PARCC). (2013). *Sample mathematics item: Grade 6 "proportion of instruments."* Accessed at www.parcconline.org/sites/parcc/files /Grade6-ProportionsofInstruments.pdf on September 8, 2014.

Pashler, H., Rohrer, D., , N. J., & Carpenter, S. K. (2007). Enhancing learning and retarding forgetting: Choices and consequences. *Psychonomic Bulletin & Review, 14*(2), 187–193.

Popham, W. J. (2011). *Transformative assessment in action: An inside look at applying the process.* Alexandria, VA: Association for Supervision and Curriculum Development.

Reeves, D. (2011). *Elements of grading: A guide to effective practice.* Bloomington, IN: Solution Tree Press.

Resnick, L. B. (Ed.). (2006). Do the math: Cognitive demand makes a difference. *Research Points: Essential Information for Education Policy, 4*(2), 1–4. Accessed at www.aera.net/Portals/38/docs /Publications/Do%20the%20Math.pdf on January 22, 2014.

Rohrer, D., & Pashler, H. (2007). Increasing retention without increasing study time. *Current Directions in Psychological Science, 16*(4), 183–186. Accessed at www.pashler.com/Articles /RohrerPashler2007CDPS.pdf on March 10, 2014.

Rohrer, D., & Pashler, H. (2010). Recent research on human learning challenges conventional instructional strategies. *Educational Researcher, 39*(5), 406–412.

Shuhua, A. (2004). *The middle path in math instruction: Solutions for improving math education.* Lanham, MD: Scarecrow Education.

Silver, E. (2010). Examining what teachers do when they display their best practice: Teaching mathematics for understanding. *Journal of Mathematics Education at Teachers College, 1*(1), 1–6.

Smarter Balanced Assessment Consortium. (n.d.). *Expressions and equations 2 sample item grade 7.* Accessed at http://sampleitems.smarterbalanced.org/itempreview/sbac/index.htm on September 8, 2014.

Smarter Balanced Assessment Consortium. (n.d.). *Sandbags 1 sample item grade 6.* Accessed at http://sampleitems.smarterbalanced.org/itempreview/sbac/index.htm on September 8, 2014.

Smarter Balanced Assessment Consortium. (n.d.). *Sandbags 2 sample item grade 6.* Accessed at http://sampleitems.smarterbalanced.org/itempreview/sbac/index.htm on September 8, 2014.

Smith, M. S., Bill, V., & Hughes, E. K. (2008). Thinking through a lesson: Successfully implementing high-level tasks. *Mathematics Teaching in the Middle School, 14*(3), 132–138.

Smith, M. S., & Stein, M. K. (1998). Selecting and creating mathematical tasks: From research to practice. *Mathematics Teaching in the Middle School, 3*(5), 348.

Smith, M. S., & Stein, M. K. (2011). *5 practices for orchestrating productive mathematics discussions.* Reston, VA: National Council of Teachers of Mathematics.

Smith, M. S., & Stein, M. K. (2012). Selecting and creating mathematical tasks: From research to practice. In G. Lappan, M. K. Smith, & E. Jones (Eds.), *Rich and engaging mathematical tasks, grades 5–9* (pp. 344–350). Reston, VA: National Council of Teachers of Mathematics.

Stein, M. K., Grover, B. W., & Henningsen, M. (1996). Building student capacity for mathematical thinking and reasoning: An analysis of mathematical tasks used in reform classrooms. *American Educational Research Journal, 33*(2), 455–488.

Stein, M. K., Remillard, J., & Smith, M. S. (2007). How curriculum influences student learning. In F. K. Lester Jr. (Ed.), *Second handbook of research on mathematics teaching and learning: A project of the National Council of Teachers of Mathematics* (pp. 319–370). Charlotte, NC: Information Age.

Wallace, W. V. (2013). *Formative assessment: Benefit for all.* Unpublished master's thesis, University of Central Florida, Orlando.

Webb, N. L. (1997). *Criteria for alignment of expectations and assessments in mathematics and science education* (Research Monograph No. 8). Washington, DC: Council of Chief State School Officers.

Webb, N. L. (2002). *Depth-of-knowledge levels for four content areas.* Accessed at www.allentownsd.org/cms/lib01/PA01001524/Centricity/Domain/1502/depth%20of%20knowledge%20guide%20for%20a!l%20subject%20areas.pdf on February 27, 2014.

Wiliam, D. (2007). Keeping learning on track: Classroom assessment and the regulation of learning. In F. K. Lester Jr. (Ed.), *Second handbook of research on mathematics teaching and learning: A project of the National Council of Teachers of Mathematics* (pp. 1051–1098). Charlotte, NC: Information Age.

Wiliam, D. (2011). *Embedded formative assessment.* Bloomington, IN: Solution Tree Press.

Index

Common Core Mathematics in a PLC at Work™ series
Edited by Timothy D. Kanold
By Thomasenia Lott Adams, Harold Asturias, Diane J. Briars, John A. Carter, Juli K. Dixon, Francis (Skip) Fennell, David Foster, Mardi A. Gale, Timothy D. Kanold, Beth McCord Kobett, Matthew R. Larson, Mona Toncheff, Jonathan A. Wray, and Gwendolyn Zimmermann

These teacher guides illustrate how to sustain successful implementation of the Common Core State Standards for mathematics. Discover what students should learn and how they should learn it at each grade level. Comprehensive and research-affirmed analysis tools and strategies will help you and your collaborative team develop and assess student demonstrations of deep conceptual understanding *and* procedural fluency.
BKF566, BKF568, BKF574, BKF561, BKF559

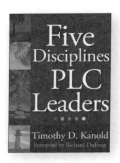

The Five Disciplines of PLC Leaders
By Timothy D. Kanold
Foreword by Richard DuFour

Outstanding leadership in a professional learning community requires practice and patience. Simply trying harder will not yield results; leaders must proactively *train* to get better at the skills that matter. This book offers a framework to focus time, energy, and effort on five key disciplines. Included are reflection exercises to help readers find their own path toward effective PLC leadership.
BKF495

Becoming an Authentic Learning Leader: Whatever You Do, Inspire Me
Featuring Timothy D. Kanold

Encourage your skeptics, cynics, and rebels with Dr. Kanold's eight fundamental disciplines of inspirational leadership. These essential concepts can impact your leadership life as well as the legacy of your leadership team. Practical yet challenging, this humorous and motivational session provides the support and focus needed to sustain effective leadership over time.
DVF063

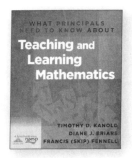

What Principals Need to Know About Teaching and Learning Mathematics
By Timothy D. Kanold, Diane J. Briars, and Francis (Skip) Fennell

Ensure a challenging mathematics experience for every learner, every day. This must-have resource offers support and encouragement for improved mathematics achievement across every grade level of your school. With an emphasis on *Principles and Standards for School Mathematics* and Common Core State Standards, this book covers the importance of mathematics content, learning and instruction, and mathematics assessment.
BKF501

Tremendous, tremendous, tremendous!

The speaker made me do some very deep internal reflection about the **PLC process** and the personal responsibility I have in making the school improvement process work **for ALL kids.**

—Marc Rodriguez, teacher effectiveness coach, Denver Public Schools, Colorado